D1121919

X-ray Diffraction by Disordered and Ordered Systems

Covering X-ray Diffraction by Gases, Liquids and Solids and Indicating How the Theory of Diffraction by These Different States of Matter is Related and How It Can Be Used to Solve Structural Problems

Related Pergamon Titles of Interest

Books

ASHBY
Dislocation Modelling of Physical Systems

BATA
Advances in Liquid Crystal Research and Applications (2 Vols)

CARTER
Metallic Shifts in NMR

GRAY & GOODBY
Liquid Crystals

HOPKINS
The Mechanics of Solids

KRISHNAN *et al*
Thermal Expansion of Crystals

PAMPLIN
Crystal Growth 2nd edition

TANNER
X-ray Diffraction Topography

WHITTAKER
Crystallography (An Introduction for Earth Science (and Other Solid State) Students)

Journals

Journal of Physics and Chemistry of Solids

Materials and Society

Materials Research Bulletin

Progress in Biophysics and Molecular Biology

Progress in Materials Science

Progress in Crystal Growth and Characterization

Solid State Communications

X-ray Diffraction by Disordered and Ordered Systems

Covering X-ray Diffraction by Gases, Liquids and Solids and Indicating How the Theory of Diffraction by These Different States of Matter is Related and How It Can Be Used to Solve Structural Problems

by

DAVID W L HUKINS

Department of Medical Biophysics
University of Manchester, U.K.

PERGAMON PRESS

OXFORD · NEW YORK · TORONTO · SYDNEY · PARIS · FRANKFURT

U.K.	Pergamon Press Ltd., Headington Hill Hall, Oxford OX3 0BW, England
U.S.A.	Pergamon Press Inc., Maxwell House, Fairview Park, Elmsford, New York 10523, U.S.A.
CANADA	Pergamon Press Canada Ltd., Suite 104, 150 Consumers Rd., Willowdale, Ontario M2J 1P9, Canada
AUSTRALIA	Pergamon Press (Aust.) Pty. Ltd., P.O. Box 544, Potts Point, N.S.W. 2011, Australia
FRANCE	Pergamon Press SARL, 24 rue des Ecoles, 75240 Paris, Cedex 05, France
FEDERAL REPUBLIC OF GERMANY	Pergamon Press GmbH, 6242 Kronberg-Taunus, Hammerweg 6, Federal Republic of Germany

First edition 1981

British Library Cataloguing in Publication Data

Hukins, David
X-ray diffraction by disordered and ordered systems.
1. X-rays—Diffraction
I. Title
548'.83 QC482.D5

ISBN 0-08-023976-5

In order to make this volume available as economically and as rapidly as possible the author's typescript has been reproduced in its original form. This method unfortunately has its typographical limitations but it is hoped that they in no way distract the reader.

Printed in Great Britain by A. Wheaton & Co. Ltd., Exeter

Preface

This book is intended primarily for those readers, especially beginning research students, who wish to use X-ray diffraction to investigate specimens which may not be crystalline. Beginners need to understand how interference effects arise in X-ray diffraction patterns and the underlying principles involved in developing structural models. A general approach to these topics, and diffraction geometry, appears in the first four chapters of the book. Next the general ideas are applied to the scattering of X-rays by isolated atoms and molecules. In subsequent chapters the ways in which interference modifies the pattern of scattered X-rays, when the atoms and molecules aggregate in various states of matter, are considered. Such states of matter include liquids, amorphous solids and liquid crystals as well as crystals. Each state gives rise to characteristic interference effects which influence the appearance and any subsequent analysis of the resulting X-ray diffraction pattern. The final chapter demonstrates that the interpretation of X-ray diffraction patterns is simply an application of physical optics.

Although readers will have a common interest in interpreting X-ray diffraction patterns, their backgrounds are likely to be very different. The technique of X-ray diffraction is used by biologists, biochemists, geologists, chemists, physicists and materials scientists. I have, therefore, tried to explain all except the most elementary physical concepts which arise. But I have usually kept these explanations fairly terse in an attempt to avoid boring readers with a strong background in physics. Similarly I have often supplemented mathematics with descriptions in words. On occasion I have avoided mathematics, such as explicit use of the Dirac delta function, where it might have distracted some readers. Nevertheless mathematical ideas are used wherever I believe they are needed for a clear and concise explanation.

My intention is that this book will also be useful to advanced undergraduates and students attending post-graduate lecture courses. Books on diffraction by crystals can be seriously misleading for these students because they often create the false impression that X-ray diffraction patterns can only be recorded from crystals. If they avoid this trap, students risk being caught by another - the misconception that there must be something resembling a lattice in systems such as liquids or gels if interference effects are to modulate the pattern of X-rays scattered by their constituent molecules. I hope that this book will help to prevent the spread of these misunderstandings.

SI units are used throughout. Units can cause confusion when comparing formulae for the intensity of X-rays scattered by an electron given in different books - some of the books listed in the bibliography use cgs units. Similar difficulties occur when relating the intensity of scattered light to molecular polarisability - cgs units are used in many books on light scattering and the formulae are different. Wavelengths and distances between atoms are measured in nanometres (1 nm = 10^{-9} m) or picometres (1 pm = 10^{-12} m).

Many colleagues and friends have influenced this book - either indirectly, by their influence on my ideas, or directly, by helping with various stages in its production. My ideas owe much to the influence of Struther Arnott who taught me the value of analysing diffraction patterns from specimens which are not single crystals. Conversations with John Woodhead-Galloway helped me to understand more about the liquid state; our collaborative research has also influenced the treatment of Section IX.6. A meeting with Alan MacKay, when the book was being planned, proved encouraging and influenced its contents. Richard Aspden read the whole book, as it was being written, and spotted many errors and ambiguities - some may remain because I have not always taken his advice. Margaret Adams, Stuart Grundy and Steve Hickey kindly provided photographs. Karen Davies helped to obtain many of the diffraction patterns as well as photographing all the other illustrations which were drawn by Percy Hukins, my father. The typescript was produced by Laurel Henfrey Business Services of Bramhall, Cheshire. Celia Hukins, my wife, helped me with the index. I am very grateful both to all of them and to the many others who have, in some way, helped to determine the final form of this book.

Contents

CHAPTER I

Scattering by an Electron

I.1. Introduction

X-Ray diffraction is used to investigate the structure of matter at the molecular level. Probably the most common application is to determine the positions of atoms in crystals. If these atoms are chemically bonded to each other, the technique yields the positions of the atoms in a molecule ie. the molecular structure. The determination of molecular structures in this way is a routine procedure in chemistry and, increasingly, in biochemistry. However, the technique of X-ray diffraction is far more versatile than this common application might suggest. Useful data can be obtained from a wide variety of states of matter and the analysis of these data is not restricted to the determination of molecular structure.

Applications include investigation of the arrangement of atoms and molecules in various phases and in sub-microscopic structures which are assemblies of molecules. Examples are provided by the investigation of the structure of liquids by physicists and of the preferred orientations of polymer molecules in fibres and films by materials scientists. Biologists use X-ray diffraction to determine the arrangements of molecules in such complicated and diverse systems as muscle, viruses and cell membranes.

The technique can also be used as a simple and rapid method for identifying unknown substances. Reference X-ray diffraction patterns are recorded from a range of samples and these are compared with the pattern from the unidentified substance. The substance is identified if its pattern coincides with one of the reference patterns. This procedure is a common analytical method for identifying the chemical composition of microcrystals in powders and is frequently used for this purpose by chemists and geologists - it is even sometimes used by pathologists who wish to identify the crystalline deposits which appear in certain diseases eg. sodium urate which is deposited in cases of gout.

X-Rays are scattered by the electrons in matter; this chapter deals with those aspects of X-ray scattering by an electron which form the foundation for the analysis of X-ray diffraction patterns from real systems. Although scattering by electrons and atoms could be regarded as a problem in quantum mechanics, the quantum nature of these phenomena will largely be ignored here - except that the shortcomings of classical theory will be noted where necessary. Classical theory provides a simple physical picture of the scattering process which is in surprisingly

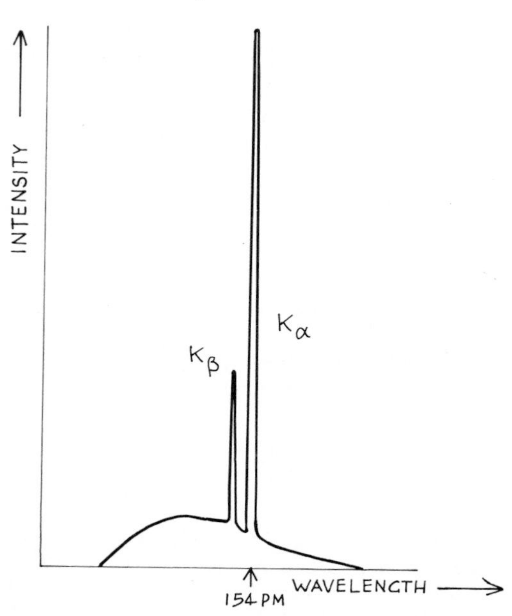

Fig. I.1. Intensity of X-rays from a typical generator as a function of wavelength; the wavelength of K_α X-rays produced with a copper anode is marked.

good agreement with quantum mechanics and, more importantly, experimental observations.

I.2. X-Ray beams

In this book the X-ray beam which is incident on a specimen will always be considered to be monochromatic and collimated; production of such a beam is reasonably simple. Figure I.1 shows the intensity of the X-rays produced by a typical generator, with a copper anode, as a function of wavelength. X-Rays emitted with the "characteristic" wavelengths, marked K_α and K_β in the figure, are considerably more intense than the radiation at other wavelengths - the so-called "bremsstrahlung". An effectively monochromatic beam can be obtained by filtering out the K_β X-rays with a sheet of a suitable metal foil - nickel in the case of X-rays generated at a copper anode. (This filter exploits the sharp discontinuity in the absorption coefficient of nickel - its "absorption edge" - which occurs at a wavelength between those of copper K_α and K_β X-rays; further details are given in Section V.5.) The resulting monochromatic beam can then be collimated by a series of pin-holes which, typically, have a diameter of around 100 μm.

Details of the apparatus used for X-ray diffraction experiments are beyond the scope of this book. Such details are already widely available and, with the exception of the cameras and diffractometers used to record diffraction patterns from single crystals, the apparatus is conceptually very simple. An idea of this conceptual simplicity can be gained from Figs. IV.4 and IV.6 of Section IV.3. The description in the preceding paragraph was intended merely to give some indication of the methods which are adopted. Suitable beams can also be obtained by focusing the X-rays using either of two different methods. One exploits the total internal

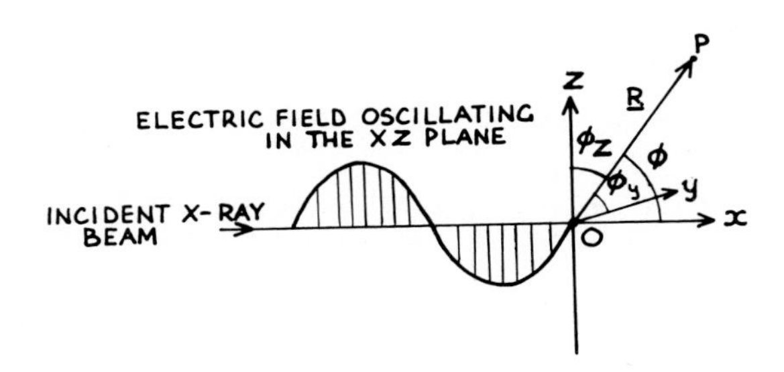

Fig. I.2. Polarised X-ray beam incident on an electron.

reflection of X-rays in air at suitable surfaces; the glancing angle at a glass surface for copper K_α X-rays is 0.25°. The other exploits "Bragg reflection" by a crystal (Section X.2). A crystal also acts as a monochromator since X-rays with different wavelengths are "reflected" at different angles; thus K_α and K_β radiation can be separated and the K_α line can even be split into its two closely spaced components - $K_{\alpha 1}$ and $K_{\alpha 2}$. If the reflector is bent, the beam can be focused.

I.3. Polarised X-rays

In Fig. I.2 a single electron defines the origin, O, of a coordinate system. A beam of plane polarised X-rays is incident on this electron and its direction of propagation defines the x-axis. Its plane of polarisation, ie. the plane in which its electric vector oscillates, defines the z-axis and the y-axis is defined so as to form a right-handed Cartesian coordinate system. The apparent ambiguity in the definition of the z-axis (should it point up or down in Fig. I.2?) is of no concern because the system being investigated has mirror symmetry in the xy plane.

Because of its electric charge, the electron oscillates up and down the z-axis with the same frequency as the electric vector of the incident X-ray beam; consequently it acts as a source of X-rays itself since accelerating charges are the source of electromagnetic radiation. If $\underline{E}$ is the electric field at the electron (where $\underline{E}$ acts up or down the z-axis), at some instant in time, the force acting on it is $e\underline{E}$ where e is the electron's charge (-1.6×10^{-19} C). This force can be equated with the mass, m, of the electron (9.1×10^{-31} kg) multiplied by its acceleration - which is, therefore, $(e/m)\underline{E}$. Since e is negative, the acceleration of the electron is in the opposite direction to $\underline{E}$; thus the oscillating electron, and hence the X-rays it emits, are π radians out of phase with the incident X-rays. The emitted X-rays will have the same frequency as the incident beam - and, therefore, the same wavelength.

The intensity of X-rays from the oscillating electron detected by an observer at P can be calculated from classical electrodynamics. If I_o is the intensity incident on the electron, the intensity detected by the observer is given, in SI units, by

$$I_s = I_o(\mu_o/4\pi)^2(e^4/m^2R^2)\sin^2\phi_z \qquad (I.1)$$

Here R is the modulus of the vector, $\underline{R}$, which defines the position of P and which makes an angle ϕ_z with the z-axis; μ_o is the permeability of free space ($4\pi \times 10^{-7}$ H m^{-1}).

Equation I.1 explains why we are able to disregard the scattering of X-rays by the protons in matter. The property of the electron which causes it to act as a secondary source, in a beam of incident X-rays, is its charge. We might expect that the proton, which has an equal but opposite charge to that of an electron, would have this same property. However, the mass of a proton (1.7×10^{-27} kg) is very considerably greater than that of an electron. Since I_s is inversely proportional to the square of the mass of the charged particle, the intensity of X-rays propagated by an oscillating proton is negligibly low when compared with that from an electron.

X-Rays are propagated by the oscillating electron in all directions; thus the electron can be considered to scatter the incident X-ray beam. These scattered X-rays are also plane polarised. It is not intended to provide any justification for this statement or to prove equation I.1. The scattering of plane polarised X-rays by an electron is the starting point from which most of the analysis in the rest of this book proceeds. Only in Section V.5 will there be any need to consider the derivation of equation I.1 any further.

I.4. Unpolarised X-rays

If the incident X-ray beam is not polarised, its electric vector can, nevertheless, be resolved into two components which are perpendicular to each other; each component then behaves like a plane polarised beam. One component defines the z-axis, as in Fig. I.2, and the other oscillates in the y-axis direction, defined in the same way as before - since they are perpendicular, they may be treated independently. Each independent component has an equal amplitude, and hence an equal intensity, if the beam is completely non-polarised. Since the overall intensity of this beam is denoted by I_o, each component has an intensity of $(I_o/2)$. Equation I.1 can then be applied to each component separately and the total scattered intensity, the sum of the two results, is given by

$$I_s = (I_o/2)(\mu_o/4\pi)^2(e^4/m^2R^2)(\sin^2\phi_z + \sin^2\phi_y) \tag{I.2}$$

where ϕ_y is the angle made by $\underline{R}$ with the y-axis as in Fig. I.2. If ϕ is the angle which $\underline{R}$ makes with the x-axis, then, from the properties of the direction cosines of P,

$$\cos^2\phi + \cos^2\phi_y + \cos^2\phi_z = 1 \tag{I.3}$$

The angle ϕ will be encountered frequently in the rest of this book because it can be used to describe the direction of scattering with respect to the direction of the incident beam. From equations I.2 and I.3, it follows that

$$I_s = I_o(\mu_o/4\pi)^2(e^4/m^2R^2)[(1 + \cos^2\phi)/2] \tag{I.4}$$

Equations I.1 and I.4 differ only in the form of a trigonometric factor which depends on the state of polarisation of the incident X-ray beam. Hence they can both be combined in the general form :

$$\left.\begin{aligned} I_s &= I_o(\mu_o/4\pi)^2(e^4/m^2R^2)p \\ p &= \begin{cases} \sin^2\phi_z & \text{, polarised} \\ (1 + \cos^2\phi)/2 & \text{, unpolarised} \end{cases} \end{aligned}\right\} \tag{I.5}$$

Equation I.5 defines the "polarisation factor", p, of X-ray diffraction analysis. A conventional X-ray generator produces unpolarised X-rays and so the lower expression for p, in equation I.5, is usually applicable. However, an unpolarised beam of X-rays may be partially plane polarised by a crystal monochromator. Synchrotrons and storage rings, which are increasingly being exploited as sources of intense X-rays, also produce plane polarised radiation; the X-rays are emitted by electrons or positrons which are constantly accelerating because they are moving in a circular path. The speed of the particles, electrons or positrons, in a synchrotron or storage ring approaches that of light.

I.5. Compton effect

According to classical electrodynamics, X-rays are scattered by electrons without their wavelength being changed; quantum theory predicts a slight complication. In order to explain quantum effects, it is necessary to think of the X-ray beam as a stream of photons. Each photon has an energy of hc/λ where h is Planck's constant (6.6×10^{-34} J s), c is the speed of electromagnetic waves in a vacuum (3.0×10^{8} m s^{-1}) and λ is the wavelength of the X-rays in the wave description. Some photons collide elastically with electrons ie. they do not lose energy - returning to the wave description, λ is unchanged by scattering. Another photon may impart some of its energy to an electron in a collision and consequently there is an increase in λ. This increase is called the Compton effect. However, it never amounts to more than a few percent of λ for the copper K_{α} X-rays usually used for X-ray diffraction experiments.

X-Rays scattered without change of wavelength are coherent ie. if an incident beam is scattered by several electrons, there is a definite phase relationship between the scattered waves; Compton scattered X-rays are incoherent. Interference cannot occur between incoherent X-rays and so, in general, it might appear necessary to determine the proportion of the incident beam scattered by each process; in practice such a rigorous approach is unnecessary.

In this book it is assumed that, in an assembly of electrons, each one scatters X-rays without a change of wavelength; physical optics then predicts that the resultant scattered beam, from some systems, contains a proportion of incoherent radiation. The results from this approach agree with those from quantum theory for λ greater than about 20 pm ie. for all wavelengths used in X-ray diffraction. It might be argued that such a treatment is unsatisfactory because it attributes some of the incoherent scatter incorrectly. Quantum theory tells us that incoherent scatter can arise from the interaction of X-rays with electrons. Classical electrodynamics predicts that this interaction gives rise only to coherent scatter - the incoherent scatter must then arise from the properties of the resulting waves. Although this latter approach may be considered unsatisfactory, it has the advantage that it is easy to apply.

The results obtained for real systems, using classical theory, do not differ from the predictions of quantum theory and are in accord with experimental observations. "Quantum theory has not made the earlier theories and techniques superfluous; rather it has brought out their limitations and defined their range of validity" (M. Born and E. Wolf, "Principles of Optics", 5th ed., Pergamon Press, 1975, p xxi). Further discussion of this topic and a detailed comparison of the quantum-mechanical and classical approaches are given in the book by James (see BIBLIOGRAPHY Section 2). It is interesting that there will be negligible incoherent scatter from the electrons in a crystal because of its regular, repetitive structure.

I.6. Summary

X-rays are scattered by the electrons in matter. Classical electrodynamics can be used to calculate the scatter from a single electron; scattered intensity depends on the intensity and state of polarisation of the incident beam. The expressions for scattered intensity, for unpolarised and plane polarised X-rays, differ only in a trigonometric factor which is usually referred to as the polarisation factor. Some of the X-rays scattered by an electron are incoherent, according to quantum theory, but this complication may be neglected in practice when dealing with real systems.

CHAPTER II

Diffraction as Fourier Transformation

II.1. Introduction

If X-ray diffraction is to be applied to understanding the structure of matter at the molecular level, it is necessary to develop the theory of Chapter I to deal with assemblies of very many electrons. The structural information provided by scattered X-rays concerns the distribution of electrons since it is the electrons in matter which scatter X-rays. So far only scattering by a single electron has been considered. When an assembly of electrons scatters X-rays, interference effects are likely to occur. It is these interference effects which lead to the features of the diffraction pattern.

II.2. Scattering by an electron density distribution

In Fig. II.1 a parallel beam of monochromatic X-rays, of wavelength λ, is incident on a body. A fraction of this incident beam will be scattered through an angle ϕ. The point O is an arbitrarily defined origin on the scattering body. Unit vectors $\hat{\underline{k}}_o$ and $\hat{\underline{k}}$ are defined by the directions of the incident beam and that scattered through an angle ϕ, respectively.

Initially consideration of scatter will be restricted to that from a volume element, $\delta\underline{r}$, at a point P; the position of P in Fig. II.1 is defined by the vector $\underline{r}$. There will be a path difference between waves scattered through ϕ by electrons at P and O of

$$\underline{r}.\hat{\underline{k}} - \underline{r}.\hat{\underline{k}}_o$$

and hence a phase difference of

$$(2\pi/\lambda)\ \underline{r}.(\hat{\underline{k}} - \hat{\underline{k}}_o)$$

A vector $\underline{Q}$ is defined by

$$\underline{Q} = (2\pi/\lambda)(\hat{\underline{k}} - \hat{\underline{k}}_o) \qquad \text{(II.1)}$$

so that the phase difference simply becomes $\underline{r}.\underline{Q}$.

Now if there were only a single electron at P the wave scattered by the volume

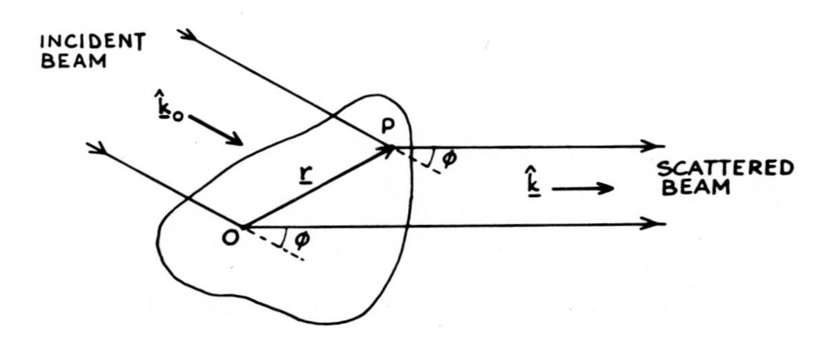

Fig. II.1. Scattering by a body.

element would have an amplitude of

$$A_o\, p^{\frac{1}{2}}\, (\mu_o/4\pi)(e^2/mR)$$

according to equation I.5 where A_o is the amplitude of the incident X-ray beam. Note that this result follows from equation I.5 because the intensity of a wave is given by the square of its amplitude. (This is not strictly true for the relationship between the modulus of the electric vector and the intensity of an electromagnetic wave. For more details of why the approach adopted here is valid see Section V.5.) If $\rho(\underline{r})$ is the electron density at P the number of electrons in the volume element will be $\rho(\underline{r})\,\delta\underline{r}$. Since interference effects are being considered, amplitudes of waves have to be added - with regard to phase differences. The electrons at P all have the same position in space and so the X-rays they scatter in a given direction will be in phase. Hence the amplitude scattered by these electrons is

$$\rho(\underline{r})\ \delta\underline{r}\ A_o\, p^{\frac{1}{2}}\, (\mu_o/4\pi)(e^2/mR)$$

Now that both the amplitude and phase of the wave scattered through an angle ϕ, by the volume element at P, are known the wave is completely specified. It may be represented by

$$\rho(\underline{r})\ \delta\underline{r}\ A_o\, p^{\frac{1}{2}}\, (\mu_o/4\pi)(e^2/mR)\ \exp\,(i\ \underline{r}.\underline{Q})$$

The resultant wave scattered through an angle ϕ by the entire body is then given by

$$F(\underline{Q}) = A_o\, p^{\frac{1}{2}}\, \frac{\mu_o}{4\pi}\, \frac{e^2}{mR} \int \rho(\underline{r})\ \exp\,(i\ \underline{r}.\underline{Q})\ \ d\underline{r} \qquad \text{(II.2)}$$

where the limits of integration are over the entire volume of the scattering body.

II.3. Representation of scattered waves

Equation II.2 incorporates the dependence of the scattered wave on a number of factors which are usually of little concern. The quantities e, m and μ_o are fundamental constants; the appearance of R simply takes into account the inverse square dependence of intensity on distance. A_o acts only as a scale factor. All

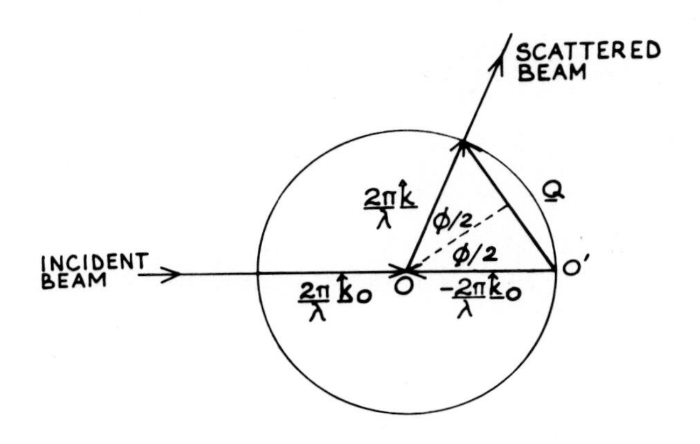

Fig. II.2. Definition of $\underline{Q}$

of these factors can be ignored because, in this book, only the relative intensities of the X-rays scattered in different directions by a body need be considered; in many applications of X-ray diffraction it is unnecessary to measure scattered intensity on an absolute scale. If the polarisation factor, p, is sufficiently different from unity, it is usual to correct the observed data to remove its effect.

Thus, for the purposes of this book, the resultant wave scattered through an angle ϕ by a body can be represented, both in amplitude and phase, by

$$F(\underline{Q}) = \int \rho(\underline{r}) \exp(i\, \underline{r}.\underline{Q})\, d\underline{r} \qquad \text{(II.3)}$$

Equation II.3 follows from equation II.2; the integration is over the whole of the scattering sample. Two parameters, $\underline{Q}$ and ϕ, have so far been used to describe the direction of the scattered beam and it is essential to relate them. Figure II.2 shows $\underline{Q}$ as defined by equation II.1. It is clear from this figure that $\underline{Q}$ defines a space whose origin is at O', which is distant $2\pi/\lambda$ from the origin, O, of "real" space, and that

$$Q = |\underline{Q}| = (4\pi/\lambda) \sin(\phi/2) \qquad \text{(II.4)}$$

If the integral of equation II.3 were evaluated it would, in general, be a complex number of the form $A \exp i\alpha$. The modulus, A, of this complex number represents the amplitude of the wave and α its phase. Note that

$$A \exp i\alpha = A \cos\alpha + i\, A \sin\alpha$$

so that the imaginary part of $F(\underline{Q})$ divided by the real part yields the tangent of the phase.

The intensity of X-rays reaching a unit area of detector, in a given time, depends on the power transmitted by a unit cross-section of the beam. For a wave which is represented by a complex number, $F(\underline{Q})$, this intensity is given by $F(\underline{Q})F^*(\underline{Q})$, where F^* is the complex conjugate of F. Thus if $I(\underline{Q})$ is the detected X-ray intensity scattered through an angle ϕ

$$I(\underline{Q}) = F(\underline{Q})\ F^*(\underline{Q}) \qquad \text{(II.5)}$$

If the integration of equation II.3 is taken over all space, then $F(\underline{Q})$ is called the Fourier transform of $\rho(\underline{r})$. When this equation was derived, the scattering body was considered to be the only source of electron density in accessible space; hence integration over the extent of the body is the same as integration over all space. Thus $F(\underline{Q})$ may be considered to be the Fourier transform of $\rho(\underline{r})$. In consequence the properties of the Fourier transform may be used to explain the properties of scattered waves.

II.4. Properties of the Fourier transform

In this section I shall simply list some properties of the Fourier transform. No proofs are given, either because these properties follow simply from the definition of the transform or because they are given elsewhere; the book by Champeney (see BIBLIOGRAPHY Section 4) is a particularly useful source of proofs. For the purposes of this section, the Fourier transform, $F(\underline{Q})$, of a function, $\rho(\underline{r})$, is defined by

$$F(\underline{Q}) = \int_s \rho(\underline{r})\ \exp\ (i\ \underline{r}.\underline{Q})\ \ d\underline{r} \qquad \text{(II.6)}$$

where s at the foot of the integral sign denotes integration over all space.

The simplest properties concern addition and rotation. If two functions are added the transform of their sum is simply the sum of the transforms of the two functions. Note that in general $F(\underline{Q})$ will be a complex number so that addition of Fourier transforms involves maintaining the correct phase relationship between them. The need to maintain this phase relationship allows the transform to be used to represent a wave. Here the addition properties of $F(\underline{Q})$ coincide exactly with the "Principle of Superposition" as described in elementary text-books dealing with wave phenomena. If a function is rotated, then an equal rotation, in the same direction about a parallel axis, has to be applied to its Fourier transform.

If equation II.6 defines $F(\underline{Q})$ as the Fourier transform of $\rho(\underline{r})$, then the inversion theorem states that

$$\rho(\underline{r}) = \int_s F(\underline{Q})\ \exp\ (-\ i\ \underline{r}.\underline{Q})\ \ d\underline{Q} \qquad \text{(II.7)}$$

The integral in equation II.7 is often termed the inverse Fourier transform, if equation II.6 defines the Fourier transform. Strictly the integral should be multiplied by $(2\pi)^{-3}$ when it applies to a three-dimensional space. In Section III.3 it was stated that, in this book, $F(\underline{Q})$ will not be considered to be measured on an absolute scale. Thus equation II.7 will not return $\rho(\underline{r})$ from $F(\underline{Q})$ on an absolute scale - so that the scale factor of $(2\pi)^{-3}$ which arises on inverse transformation will be neglected in the rest of this book. It makes no difference which of equation II.6 or II.7 is the form of integral defining the Fourier transform since the other automatically becomes the inverse transform. Both equations II.6 and II.7 have the same properties and for many purposes it does not matter which is used. The two forms of integral differ only in the sign of their phases. Thus $\rho'(\underline{r})$ calculated from

$$\rho'(\underline{r}) = \int_s F(\underline{Q})\ \exp\ (i\ \underline{r}.\underline{Q})\ \ d\underline{Q}$$

differs from $\rho(\underline{r})$, calculated using equation II.7, only in that one is inverted in the origin with respect to the other.

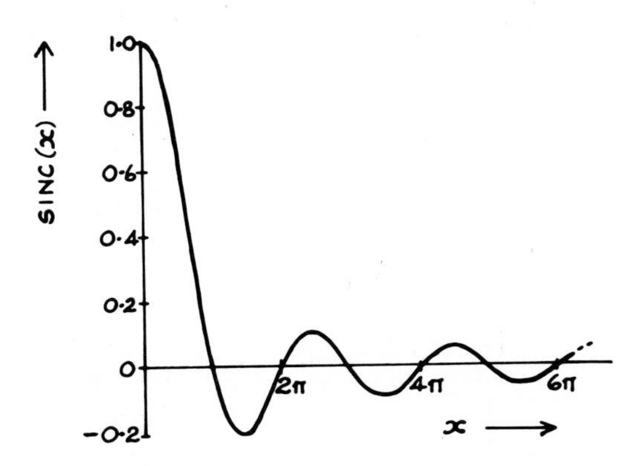

Fig. II.3. The function sinc x.

Suppose $\rho(\underline{r})$ depends only on r, the modulus of $\underline{r}$, and is independent of its direction. Since most of the applications described in this book will involve three-dimensional examples of $\rho(\underline{r})$, it will then have spherical symmetry. This spherically symmetric electron density can be represented by $\rho(r)$ and it turns out that $F(\underline{Q})$ depends only on the modulus of $\underline{Q}$ and not on its direction. $F(\underline{Q})$, the spherically symmetric Fourier transform, is then given by

$$\left.\begin{aligned} F(Q) &= \int_0^\infty 4\pi r^2 \rho(r) \operatorname{sinc}(Qr)\, dr \\ \operatorname{sinc} x &= x^{-1} \sin x \end{aligned}\right\} \quad \text{(II.8)}$$

The function sinc x is plotted in Fig. II.3; its values can be either positive or negative. Note that F(Q) is no longer complex but it still conveys phase information. In Section II.3 $F(\underline{Q})$ was represented in the form

$$A \cos \alpha + i A \sin \alpha$$

but here F(Q) is real so that

$$A \sin \alpha = 0$$

This condition is only met if α is given by

$$\alpha = n\pi \qquad n = 0, \pm 1, \pm 2 - - - -$$

Since phases are measured by the displacement of one wave relative to another, a real value of F(Q) then implies a phase of zero or π radians. (Throughout this book phases will be measured with respect to a wave scattered with a zero value of Q.) If the phase is zero then A cos α is positive and hence F(Q) is positive; if the phase is π then A cos α is negative and hence F(Q) is negative. Thus, for a spherically symmetric $\rho(\underline{r})$, F(Q) will always be real and can be positive (corresponding to a phase of zero for the wave scattered in the direction specified by Q) or negative (corresponding to a phase of π).

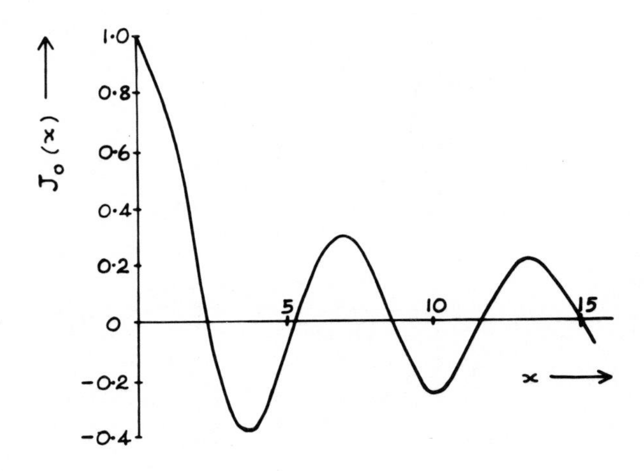

Fig. II.4. The function $J_o(x)$.

In Chapter IX some cylindrically symmetric systems will be encountered; in these systems $\rho(\underline{r})$ is independent of the direction of $\underline{r}$ in projection. The Fourier transform of these two-dimensional projections also depends only on the modulus of $\underline{Q}$ and not on its direction. $F(Q)$, the cylindrically symmetric Fourier transform is then given by

$$F(Q) = \int_o^{\infty} 2\pi r \ \rho(r) \ J_o(Qr) \ dr \tag{II.9}$$

J_o, in equation II.9, is the zero-order Bessel function of the first kind and is plotted in Fig. II.4. The Bessel function of the first kind, of order n, may be defined by

$$J_n(x) = \sum_{m=0}^{\infty} \frac{(-1)^m \ (x/2)^{n+2m}}{m! \ (n + m)!} \tag{II.10}$$

In general the series of equation II.10 does not converge rapidly and this function is usually computed using a recurrence relation. Further information on Bessel functions appears in Section IX.2 and in the book by Watson (see BIBLIOGRAPHY Section 4). Before continuing, it is perhaps worth noting that, just as a spherically symmetric function could be considered as a series of concentric spherical shells, so a cylindrically symmetric function could be considered as a series of coaxial cylindrical shells.

If two functions are multiplied, then the Fourier transform of their product is given by the convolution of the Fourier transforms of the two functions. The convolution of two functions $\rho(\underline{r})$ and $\psi(\underline{r})$ is written $\rho(\underline{r}) \otimes \psi(\underline{r})$ and is defined by

$$\rho(\underline{r}) \otimes \psi(\underline{r}) = \int_s \rho(\underline{r}_1) \ \psi(\underline{r} - \underline{r}_1) \ d\underline{r}_1 \tag{II.11}$$

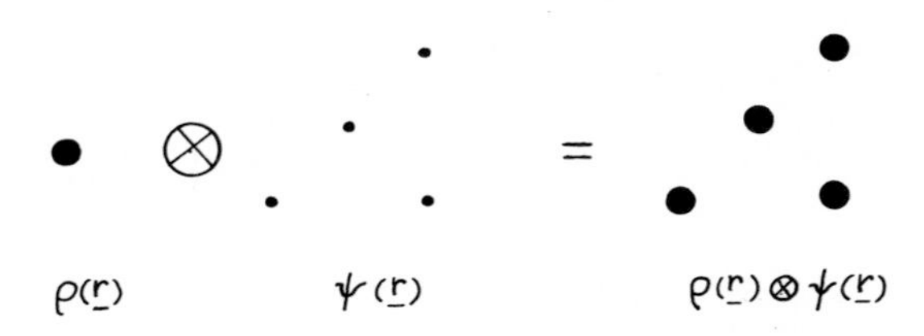

Fig. II.5. Two-dimensional example of the convolution of a function with a set of points.

Thus $\rho(\underline{r})$ is smeared out in a way which is dictated by the form of $\psi(\underline{r})$. Equation II.11 is particularly easy to understand physically if $\psi(\underline{r})$ is non-zero only at a set of points. Figure II.5 shows a two-dimensional example where the convolution operation consists simply of setting down $\rho(\underline{r})$ with its origin at each of the points.

II.5. Light

This section and the next are digressions from the main theme of the book and may be omitted without loss of continuity; here the application of X-ray diffraction theory to the diffraction of light will be discussed. Since light is simply electromagnetic radiation with a longer wavelength (0.4 to 0.7 μm) than X-rays, it might appear that a separate section on its properties is unnecessary. However there are two reasons for considering the behaviour of light in some detail. Firstly, interference effects observed when light is scattered by macromolecules and colloidal particles are commonly exploited in polymer and colloid science but the theory is often presented as if it were unrelated to X-ray diffraction. Secondly, for scattering bodies whose dimensions are much greater than the wavelength, λ, of the scattered radiation, the theory breaks down for light but not for X-rays.

Since λ for light is of the order of 10^3 times greater than that of X-rays there will be no appreciable interference effects for scattering bodies which are as large as small molecules ie. which contain several electrons. Therefore, equation I.5 may be written in the form

$$I_s = I_o \; c^4 \mu_o^2 \; (\pi^2 \alpha^2/\lambda^4 R^2) \; p \tag{II.12}$$

where p is the polarisation factor and R is the distance of the detector from the scatterer as in Chapter I. Here μ_o and c also have the same meaning as in Chapter I; they are both fundamental constants whose values are given in Sections I.3 and I.5. (Note that most text-books which deal with light scattering do not use SI units and equation II.12 is then slightly different.) Molecular polarisability, α, relates the dipole moment, $\underline{P}$, induced in a molecule to the applied electric field, $\underline{E}$, and is defined by

$$\alpha\underline{E} = \underline{P}$$

The phenomenon represented by equation II.12, ie. scattering without interference, is often called "Rayleigh scattering" after Lord Rayleigh who published accounts of its theory in the late nineteenth and early twentieth centuries.

For larger molecules and colloidal particles, interference will occur between light waves scattered in the same direction by different parts of the same body. Interference effects may be calculated by the method of Section II.2; scattering from a volume element, $\delta\underline{r}$, is considered first. If V is the volume of the scattering body, the molecular polarisability of the element will be

$$\alpha\ \delta\underline{r}/V$$

if α does not vary over the body and, from equation II.12, the amplitude of the wave it scatters in a given direction is

$$A_o\ p^{\frac{1}{2}}\ (c^2\mu_o)[\pi\alpha\ \delta\underline{r}/\lambda^2\ RV]$$

where A_o is the incident amplitude. By analogy with equation II.2, the resultant wave scattered by the entire body through an angle ϕ will be represented, both in amplitude and phase, by

$$F(\underline{Q}) = A_o\ p^{\frac{1}{2}}\ (c^2\mu_o)\ \frac{\pi\alpha}{\lambda^2 RV} \int \exp\ (i\ \underline{r}.\underline{Q})\ \ d\underline{r} \qquad \text{(II.13)}$$

where Q and ϕ are related by equation II.4. Here, as before, the integration is to be taken over the entire scattering body - positions within the body are represented by $\underline{r}$. When applied to light, this approach is called the Rayleigh-Debye-Gans theory (often abbreviated to R.D.G. theory in the literature of colloid science).

Although this theory is effectively exact for X-rays it is an approximation for light scattering. In Fig. II.1, light scattered through an angle ϕ by different parts of the body will have to pass through differing thicknesses of it. Suppose that for two points the difference in thickness is t, then there will be an optical path difference of

$$t(n - n_o)$$

between the light scattered by these points, where n and n_o are the refractive indices of the body and the surrounding medium. This path difference leads to a phase difference of

$$(2\pi t/\lambda)(n - n_o)$$

which has been ignored until now. For X-rays both n and n_o will be so close to unity that this phase difference will be negligible. But for light equation II.12 is at best an approximation. The approximation is only adequate if n is not much greater than n_o and the scattering body is not too large. Exact theories of light scattering have only been worked out for a few simple shapes; details are given in the book by Kerker (see BIBLIOGRAPHY Section 4).

II.6. Neutrons and electrons

Neutron and electron diffraction can be used as alternatives to X-ray diffraction for investigating the structure of matter at the molecular level; the aim of this section is to indicate how the theory developed so far can be applied to these techniques. Particles of mass m, moving with a speed v, can be considered as a wave whose wavelength is given by

$$\lambda = h/(mv) \qquad (II.14)$$

where h is Planck's constant (whose value is given in Section I.5). Neutrons and electrons can be made to attain speeds leading to wavelengths which are comparable to those of X-rays.

Neutrons are scattered by atomic nuclei - usually elastically ie. without loss of energy. The intensity of neutrons scattered by a nucleus bears no simple relationship to its atomic number and is not the same for different isotopes of the same element. In contrast the intensity of X-rays scattered by an atom increases with the number of electrons ie. with atomic number; also the scattered intensity is the same for different isotopes of the same element because they have the same number of electrons. Neutron diffraction is thus complementary to X-ray diffraction - a particularly common application is the accurate determination of hydrogen atom positions in molecules since these atoms contain only a single electron and consequently are very weak X-ray scatterers.

A nuclear reactor is a suitably intense source of neutrons for diffraction experiments; facilities for neutron diffraction are available at reactors in England, France and the U.S.A. Each degree of translational freedom of the neutron is associated with a most probable energy of $k_BT/2$ where k_B is Boltzmann's constant (1.4×10^{-23} J K^{-1}) and T is a temperature measured on the Kelvin scale. Since the neutron has three translational degrees of freedom its kinetic energy is given by

$$mv^2/2 = 3k_BT/2 \qquad (II.15)$$

where now v is the root-mean-square velocity and m is the mass of the neutron (1.7×10^{-27} kg). From equations II.14 and II.15, the wavelength of a neutron beam is given by

$$\lambda = h/(3k_BT\ m)^{\frac{1}{2}}$$

The temperature of the neutrons from a reactor is typically of the order of 100^oC; thus λ is of the order of 0.1 nm.

Electrons are scattered by electrostatic fields in the matter of the scattering body. A suitable beam is provided by accelerating the electrons through a potential difference, V; they then acquire kinetic energy of

$$mv^2/2 = Ve \qquad (II.16)$$

where e and m are the charge and rest mass of an electron as in Section I.3. From equations II.14 and II.16 the wavelength of an electron beam is

$$\lambda = h/(2Vem)^{\frac{1}{2}} \qquad (II.17)$$

If the electron is accelerated through 10 kV, its wavelength is of the order of 0.1 nm. In practice an electron microscope is the usual source of electrons for diffraction experiments; further details are given in Section XII.4. Typically this instrument accelerates electrons through 100 kV and so their wavelength is short compared with eg. copper K_α X-rays.

The methods of this chapter are applicable to weak, elastic scattering of neutrons and electrons. Equation II.3 then takes the general form

$$F(\underline{Q}) = \int V(\underline{r})\ \exp\ (i\ \underline{r}.\underline{Q})\ \ d\underline{r} \qquad (II.18)$$

where $V(\underline{r})$ is some appropriate scattering potential function. Comparison of equations II.2 and II.3 shows that the latter neglects the fundamental constants which place $F(\underline{Q})$ on an absolute scale. A similar neglect is involved in equation II.18; a different set of constants has to be used for X-rays, neutrons and electrons because the underlying physics of the scattering process is different in each case.

Equation II.18 represents an approximate theory which is only adequate for weak scattering - the Born approximation. This approximation is analogous to the Rayleigh-Debye-Gans theory of light scattering (Section II.5). The R.D.G. theory is a reasonable approximation for small scatterers when n is not very different from n_o. Light scattering intensity is related to the molecular polarisability, α, of the scatterer which in turn is related to n, for molecules in the gas phase, by

$$\alpha = (n^2 - 1)/N$$

where N is the number of scatterers per unit volume. Hence when n is not very different from n_o the scattering is weak and the R.D.G. theory is seen to provide an example of the Born approximation for weak scattering.

II.7. Summary

Interference effects occur between X-rays scattered in the same direction by different electrons in a scattering body. The scattering direction is defined by a vector $\underline{Q}$ whose modulus is a function of scattering angle. The resultant beam scattered with a given $\underline{Q}$ is represented, both in amplitude and phase, by $F(\underline{Q})$ - the Fourier transform of the electron density, $\rho(\underline{r})$. When this beam is detected, its intensity at the detector is given by $F(\underline{Q})F^*(\underline{Q})$ where F^* is the complex conjugate of F. Recognition that the scattered waves may be represented by a Fourier transform allows the properties of this function to be used to explain diffraction effects.

CHAPTER III

Principles of Structure Determination

III.1. Introduction

The previous chapter described how to calculate the resultant X-ray wave scattered in a particular direction from the electron density distribution of a scattering body. However the only means available for investigating this electron density distribution, $\rho(\underline{r})$, is by the way it scatters radiation. Thus the aim of an X-ray diffraction experiment is to measure $F(\underline{Q})$, which represents the scattered wave, over a space defined by many $\underline{Q}$ vectors, and hence to obtain information about $\rho(\underline{r})$.

Thus the previous chapter considered the technique in reverse. In all applications of X-ray diffraction one seeks to gain information about $\rho(\underline{r})$ from $F(\underline{Q})$; the aim of the present chapter is to investigate the problems which then arise. These problems will be considered very generally without undue regard either to the particular states of matter which might be investigated in an X-ray diffraction experiment or to any specific applications.

X-Ray diffraction experiments may be performed either to measure $\rho(\underline{r})$ or to answer some restricted set of questions about its form; this chapter is concerned predominantly with the first kind of application. Measuring $\rho(\underline{r})$ amounts to determining a structure. High resolution measurement of $\rho(\underline{r})$ gives positions of atoms eg. in a molecule. At lower resolution $\rho(\underline{r})$ shows the positions of molecules. X-Ray diffraction can answer more restricted kinds of questions - two examples follow. One: are the polymer molecules in a resin oriented (is $\rho(\underline{r})$ anisotropic)? Two: does a sample of crystals consist of eg. sodium urate (is $\rho(\underline{r})$ the same as in sodium urate crystals)? Some indication of the variety of applications appeared in Section I.1 - but it may not always be clear how to categorise a problem. Nevertheless some restricted questions may be more easily answered than this chapter might imply; answering such questions usually depends on understanding the diffraction properties of particular systems as described in Chapters VI to XI.

III.2. Phase problem

At first sight it might appear that the inversion property of the Fourier transform, described in Section II.4, provides the key to the problem of calculating $\rho(\underline{r})$ from $F(\underline{Q})$. Given the definition of $F(\underline{Q})$ in equation II.6, it follows from this property that

$$\rho(\underline{r}) = \int_S F(\underline{Q}) \exp(-i\, \underline{r}.\underline{Q})\; d\underline{Q} \qquad \text{(III.1)}$$

Equation III.1 is identical to equation II.7 but is reproduced here because it is essential to the arguments of this chapter.

A problem arises in that only the modulus of $F(\underline{Q})$ can be obtained from a diffraction experiment and, in consequence, equation III.1 cannot be applied directly to calculate $\rho(\underline{r})$. The reason for this difficulty is that an X-ray detector, be it a photographic film or a counter, measures the intensity, $I(\underline{Q})$, as defined by equation II.5. In general $F(\underline{Q})$ will be complex and measurement of $I(\underline{Q})$ does not allow its real and imaginary parts to be separated. Even when $F(\underline{Q})$ is real it is necessary to know whether it is positive or negative to apply equation III.1 (see Section II.4) but this information is lost when $I(\underline{Q})$ is measured.

Since only the modulus of $F(\underline{Q})$ can be measured, it is only the amplitude and not the phase of the resultant X-ray wave scattered in the direction specified by $\underline{Q}$ which can be obtained. Hence the phases of the scattered X-rays are lost when a diffraction pattern is recorded. This loss of phase information constitutes the so-called "phase problem" of X-ray diffraction analysis.

What is the physical reason for the phase problem? For X-rays with a wavelength λ = 154 pm the time period of oscillation of the electric vector is 5×10^{-19} s. A detector of practical sensitivity will then obtain information over a time covering very many oscillations. It will therefore be insensitive to when the waves arrive at the various points over its surface and therefore to the relative phases of the resultant waves scattered in different directions. The only way to measure these phases would be by interference with a reference beam. Such an interference method requires a coherent X-ray source, eg. an X-ray laser, but such a source has yet to be developed.

Determination of a structure by X-ray diffraction thus amounts to overcoming the phase problem so that, at least in principle, $\rho(\underline{r})$ could be calculated from the completely specified $F(\underline{Q})$. There are two approaches to overcoming this problem and hence solving structures: deduction or trial-and-error. Before each of these approaches is considered in detail it is worth describing the properties of a useful function - the Patterson function.

III.3. Patterson function

The cross-correlation of two real functions $\rho(\underline{r})$ and $\psi(\underline{r})$ is denoted by $\star$ and defined by

$$\rho(\underline{r}) \star \psi(\underline{r}) = \int_S \rho(\underline{r}_1)\, \psi(\underline{r} + \underline{r}_1)\; d\underline{r}_1 \qquad \text{(III.2)}$$

Note the similarity to convolution as defined by equation II.11. The only difference is that, in the convolution operation, $\psi(\underline{r} + \underline{r}_1)$ is replaced by $\psi(\underline{r} - \underline{r}_1)$. The two operations are of course equivalent if $\psi(\underline{r} + \underline{r}_1)$ and $\psi(\underline{r} - \underline{r}_1)$ are the same ie. if $\psi(\underline{r})$ is an even function or can be transformed to one simply by translation along the direction of $\underline{r}_1$.

By analogy with equation III.2, the autocorrelation of $\rho(\underline{r})$ is defined by

$$\rho(\underline{r}) * \rho(\underline{r}) = \int_S \rho(\underline{r}_1)\ \rho(\underline{r} + \underline{r}_1)\ d\underline{r}_1 \qquad \text{(III.3)}$$

This operation produces a function which has high values whenever high values of $\rho(\underline{r}_1)$ and $\rho(\underline{r} + \underline{r}_1)$ coincide. Highest peaks occur when complete overlap of the two functions occurs. Thus the operation measures the correlation between a function and itself translated. If $\rho(\underline{r})$ repeats itself regularly it will coincide with itself, with the repeat periodicity, when translated past itself. Thus autocorrelation provides a sensitive method for the analysis of periodicities in pictures - such as electron micrographs - where $\rho(\underline{r})$ in equation III.3 represents the two-dimensional distribution of grey levels in the "black and white" picture. (It is perhaps surprising that equation II.6, which involves more computational steps than equation III.3, is so frequently used for this purpose by electron microscopists.)

A successful X-ray diffraction experiment always yields $I(\underline{Q})$ even if $F(\underline{Q})$ cannot be completely specified. What happens if $I(\underline{Q})$, which can be measured, is substituted for $F(\underline{Q})$, which generally cannot be, in equation III.1? The resulting integral is usually called the Patterson function, $P(\underline{r})$, and is defined by equation III.4 :

$$P(\underline{r}) = \int_S I(\underline{Q})\ \exp\ (-\ i\ \underline{r}.\underline{Q})\ \ d\underline{Q} \qquad \text{(III.4)}$$

$$= \int_S F(\underline{Q})\ F^*(\underline{Q})\ \exp\ (-\ i\ \underline{r}.\underline{Q})\ \ d\underline{Q} \qquad \text{[eqn.II.5]}$$

$$= \rho(\underline{r}) * \rho(\underline{r})$$

The final line is a property of the inverse transform of $F(\underline{Q})F^*(\underline{Q})$ known as the Wiener-Khintchine theorem which is proved in the book by Champeney (see BIBLIOGRAPHY Section 4). Thus the Patterson function yields the autocorrelation of the electron density in the scatterer.

The Patterson function is particularly useful when a structure contains a few atoms of high electron density. Consider what happens when $\rho(\underline{r})$ is translated past itself. Any peaks which arise from the coincidence of one of these atomic positions with another will dominate the autocorrelation function because the peak height will then be given by the product of their electron densities. Hence the autocorrelation function will be dominated by periodicities arising from the positions of the electron-dense atoms ie. the Patterson function essentially yields a map of vectors between any electron-dense atoms. Thus the relative positions of these atoms in a molecule may be deduced (using the term "molecule" in the general sense of an assembly of atoms as in Section V.3).

III.4. Solution by deduction

If $\rho(\underline{r})$ is to be deduced from an X-ray diffraction pattern, some method is required to deduce $F(\underline{Q})$ from $I(\underline{Q})$. In general this deduction will involve finding the real and imaginary parts of the complex $F(\underline{Q})$. Two different approaches have been adopted and both are conventionally described in considerable detail in textbooks of X-ray crystallography such as those recommended in Section 12 of BIBLIOGRAPHY. The aim of this section is merely to complement these standard accounts by drawing attention to how these methods relate to the principles described in

TABLE III.1. Algorithm for the "heavy atom" method

1. Collect experimental data, $I(\underline{Q})$, over a range of Q values.
2. Calculate $P(\underline{r})$ from $I(\underline{Q})$ using equation III.4.
3. Deduce positions of electron-dense atoms from $P(\underline{r})$.
4. Use positions of electron-dense atoms as first approximation to $\rho(\underline{r})$.
5. Define an initial vector $\underline{Q}$.
6. Calculate $F(\underline{Q})$ from $\rho(\underline{r})$ using equation II.3.
7. Calculate phase, $\alpha(\underline{Q})$, of the complex $F(\underline{Q})$.
8. Calculate better approximation to the complex $F(\underline{Q})$ from

$$F(\underline{Q}) = I^{\frac{1}{2}}(\underline{Q}) \exp [i\ \alpha(\underline{Q})]$$

9. Increment $\underline{Q}$.
10. If $\underline{Q}$ within range of experimental data go to 6.
11. Calculate better approximation to $\rho(\underline{r})$ from $F(\underline{Q})$ using equation III.1.
12. If better approximation required go to 5.

Chapter II.

The first approach requires the presence of a few electron-dense atoms in a molecule. Their positions can be located by the Patterson function described in Section III.3. Since such electron-dense atoms make an important contribution to $\rho(\underline{r})$ they tend, in favourable cases, to dominate the X-ray scattering and their positions can then be used as a first step in an iterative solution of the phase problem (Table III.1) which is called the "heavy atom" method. The "isomorphous replacement" method can be used if electron-dense atoms can be attached to a specimen without disrupting its structure. X-Ray diffraction experiments performed on both the specimen and this chemical derivative can then be used to solve the phase problem (Table III.2 and Fig. III.1). Both "heavy atom" and "isomorphous replacement" methods are described in considerable detail in crystallographic texts and there is little need to discuss them further here.

The second deductive approach is to recover the missing phase information by so-called "direct methods". Direct methods are illustrated here by a simple one-dimensional example which shows how $F(\underline{Q})$ can be completely deduced from $I(\underline{Q})$ in favourable circumstances. The electron density distribution along a line can be represented by $\rho(r)$ - since a point on the line can be defined, with respect to some origin, by a scalar r. From equation II.6 the Fourier transform of $\rho(r)$ is

$$F(Q) = \int_{-\infty}^{\infty} \rho(r) \exp(i\ r\ Q)\ dr$$

which, for our present purpose, can usefully be written in the form

TABLE III.2. Algorithm for the "isomorphous replacement" method

The ambiguity in step 7 can be resolved if the whole process is repeated with a second derivative. Each derivative yields two solutions; the one which appears in both cases is the required result.

1. Collect experimental data, $I(\underline{Q})$ from specimen, over a range of Q values.
2. Collect corresponding data, $I_D(\underline{Q})$, from derivative.
3. Calculate $P(\underline{r})$ from $I_D(\underline{Q})$ using equation III.4.
4. Deduce positions of electron-dense atoms in derivative from $P(\underline{r})$; denote the contribution of these atoms to the electron density of the derivative by $\rho_H(\underline{r})$.
5. Define an initial vector $\underline{Q}$.
6. Calculate $F_H(\underline{Q})$, the contribution of the electron-dense atoms to the scatter from the derivative, from $\rho_H(\underline{r})$ using equation II.3.
7. Noting that the complex $F_D(\underline{Q})$ is given by

   $$F_D(\underline{Q}) = F(\underline{Q}) + F_H(\underline{Q})$$

   calculate the complex $F(\underline{Q})$ from $F_H(\underline{Q})$, $I^{\frac{1}{2}}(\underline{Q})$ and $I_D^{\frac{1}{2}}(\underline{Q})$ as in Fig. III.1.
8. Increment $\underline{Q}$.
9. If $\underline{Q}$ still within range of experimental data go to 6.

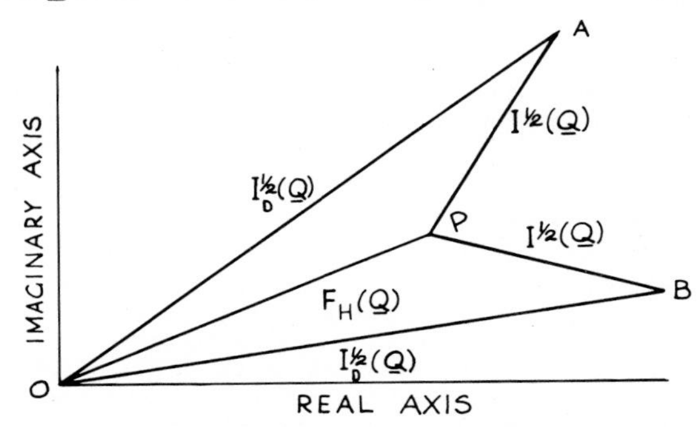

Fig. III.1. Determination of $F(\underline{Q})$, by the method of isomorphous replacement, shown on an Argand diagram. $F_H(\underline{Q})$ is represented both in amplitude and phase by OP. A circle of radius $I_D^{\frac{1}{2}}(\underline{Q})$ is drawn with O as origin. Another circle of radius $I^{\frac{1}{2}}(\underline{Q})$ is drawn with P as origin. These circles intersect at A and B. Thus either PA or PB could represent $F(\underline{Q})$ in both amplitude and phase by satisfying the relationship

$$F_D(\underline{Q}) = F(\underline{Q}) + F_H(\underline{Q})$$

$$F(Q) = \int_{-\infty}^{0} \rho(r) \exp(i r Q) \, dr + \int_{0}^{\infty} \rho(r) \exp(i r Q) \, dr$$

$$= \int_{0}^{\infty} \rho(-r) \exp(-i r Q) \, dr + \int_{0}^{\infty} \rho(r) \exp(i r Q) \, dr \qquad \text{(III.5)}$$

If $\rho(r)$ has a centre of symmetry $F(Q)$ is real; the problem of recovering the lost phase information then reduces to finding wher $F(Q)$ is positive and when it is negative - according to Section II.4. For every feature in a centrosymmetric $\rho(r)$ there will be an identical feature on the other side of the centre of symmetry. If the centre of symmetry is chosen as the origin, $\rho(r)$ becomes an even function ie. $\rho(r)$ is identical to $\rho(-r)$. Equation III.5 then becomes

$$F(Q) = \int_{0}^{\infty} \rho(r) \exp(-i r Q) \, dr + \int_{0}^{\infty} \rho(r) \exp(i r Q) \, dr$$

$$= 2 \int_{0}^{\infty} \rho(r) \cos(r Q) \, dr \qquad \text{(III.6)}$$

The final form of equation III.6 follows from the identity

$$\exp(i X) = \cos X + i \sin X$$

According to equation III.6, $F(Q)$ is real when $\rho(r)$ is centrosymmetric. The Fourier transform is also real for centrosymmetric functions in two and three dimensions; the Fourier transform of the spherically symmetric function in Section II.4 demonstrates a special case of this property. It was also explained in Section II.4 that a positive value for $F(Q)$ implies a phase of zero for the scattered X-rays and a negative value implies a phase of π radians.

Figure III.2 shows the one-dimensional $I(Q)$ - how can it be used to deduce the sign of $F(Q)$? The first peak in $I(Q)$, which occurs when Q equals zero, must arise from a positive value of $F(Q)$. The reason is, from equation II.6, that

$$F(0) = 2 \int_{0}^{\infty} \rho(r) \, dr$$

which is positive because it represents the number of scattering electrons. Suppose that $F(Q)$ is a smoothly varying function which never has a value of zero at any of its minima - which is the case for the Fourier transforms of many simple functions. When $I(Q)$ is zero, $F(Q)$ must pass through the Q-axis ie. it changes sign. Thus the second peak in $I(Q)$ must arise from a negative value of $F(Q)$. If $F(Q)$ changes sign every time $I(Q)$ passes through zero its form can be deduced with the result shown in Fig. III.3; the phases corresponding to the peaks and troughs are marked. Phases are defined with respect to a beam scattered through a zero angle which, since $F(0)$ is positive, is defined to be zero.

Direct methods are not confined to one-dimensional or centrosymmetric cases; the example here was chosen so that it would be easy to illustrate and explain. The methods are most simply applied to centrosymmetric systems but they have been extended to diffraction data from single crystals where this condition does not apply. Once again further details are given in text-books on X-ray crystallography.

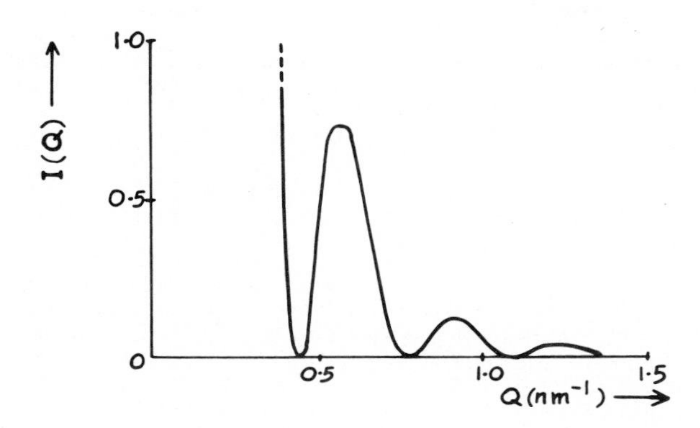

Fig. III.2. An observed intensity distribution I(Q)

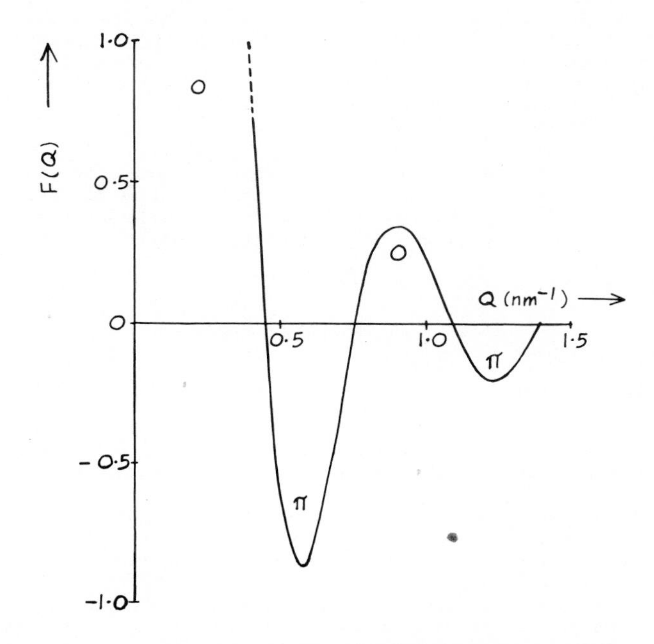

Fig. III.3. The Fourier transform, F(Q), inferred from the observed intensity distribution, I(Q), of Fig. III.2. Phases of peaks are marked.

The problem with all deductive methods of solving the phase problem is that they can only be applied in favourable cases. Usually these favourable cases are the diffraction patterns from near-perfect, three-dimensional crystals. For such systems the phase problem is scarcely a problem at all - as is demonstrated by the routine determination of crystal structures. In Chapter VI we shall see that the deductive approach can also be applied to investigating the structures of molecules in ideal gases and solutions - but here only the spherically averaged electron density can be deduced because of the nature of the specimen. Otherwise it is only in a very few special cases that X-ray diffraction data can be interpreted by deduction.

III.5. Trial-and-error solution

The first step here is to infer a trial model for $\rho(\underline{r})$ by inspection of $I(\underline{Q})$. This approach to the problem clearly requires some knowledge of the forms of the Fourier transforms of functions which might resemble $\rho(\underline{r})$, if it is to be successful. Calculation of the Patterson function from $I(\underline{Q})$ may also help in the development of a trial model. From the trial model for $\rho(\underline{r})$ the expected $I(\underline{Q})$ can be calculated using equations II.3 and II.5. The observed and calculated $I(\underline{Q})$ are then compared. On the basis of this comparison three different conclusions might be reached. One: reject the model for $\rho(\underline{r})$ as being unsatisfactory. Two: accept the model. Three: modify the model to improve the agreement.

If the trial model is not rejected outright the third possible course of action nearly always has to be adopted. A new calculation of $I(\underline{Q})$ has then to follow and the comparison repeated. Eventually, by a process of iteration, the model will be improved enough that the comparison is sufficiently favourable for the model to be accepted. This iterative process is termed "refinement". Often several trial models will have to be considered. Those that cannot be rejected outright have to be refined and the one which gives the best account of the observed $I(\underline{Q})$ accepted.

The unsatisfactory feature of trial-and-error solution is that one cannot be sure that all possible models have been considered. Thus the true structure may have been overlooked. In consequence the resulting models cannot be established with the same degree of certainty as can deductive models.

III.6. Resolution

It follows from equation II.4 that the maximum value of Q for which $F(\underline{Q})$ can be measured is given by

$$Q_{max} = 4\pi/\lambda$$

In practice it may not prove convenient, or even possible, to measure $F(\underline{Q})$ for Q values which are so great as even to approach Q_{max}. According to equation III.1, in order to calculate $\rho(\underline{r})$ from $F(\underline{Q})$ it is necessary to integrate over all the space defined by $\underline{Q}$ vectors whose moduli range between minus and plus infinity. Integration between these limits is clearly impossible because of the limited extent of the space over which $F(\underline{Q})$ values can be measured.

What is the effect of truncating the range of integration in equation III.1? The answer to this question will be illustrated by a simple one-dimensional example. Figure III.4 shows a square peak whose height, as a function of distance, represents $\rho(r)$: Fig. III.5 shows its Fourier transform, $F(Q)$. (Note that, since $\rho(r)$ here is centrosymmetric, $F(Q)$ is real and can be plotted on a single graph.) Application of equation III.1 to $F(Q)$ but with Q restricted to the range of -20 to

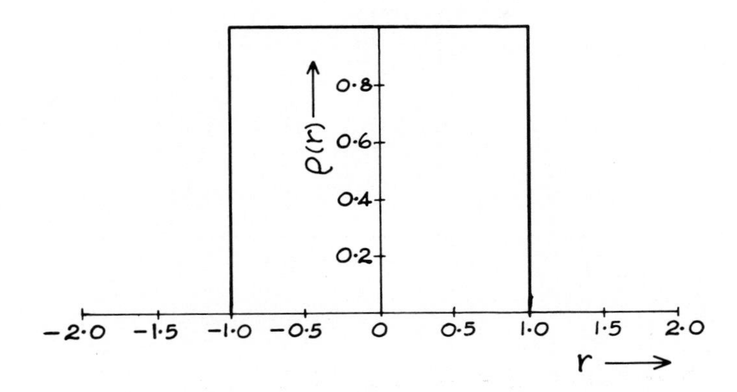

Fig. III.4. Square peak.

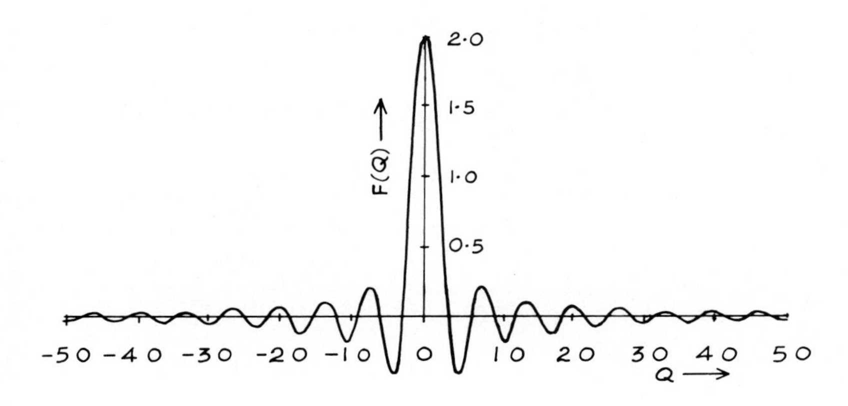

Fig. III.5. Fourier transform, F(Q), of the square peak in Fig. III.4.

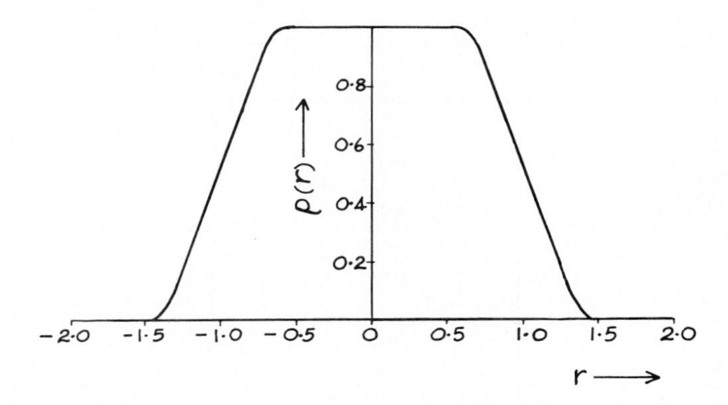

Fig. III.6. Square peak reconstructed from F(Q) of Fig. III.5 with Q in the range -5 to 5 reciprocal length units.

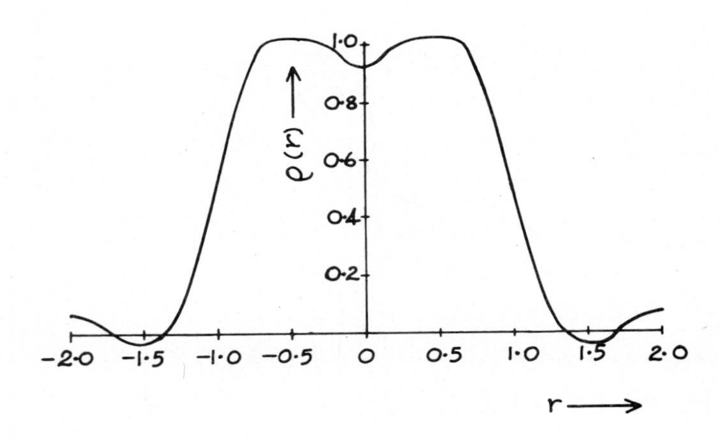

Fig. III.7. Square peak reconstructed from F(Q) of Fig. III.5 with Q in the range -6 to 6 reciprocal length units

+20 units essentially recovers the original $\rho(r)$.

As the extent of Q-space used to reconstruct $\rho(r)$ is decreased so the sharp square peak becomes less well defined. In Fig. III.6 the range of Q-space is so limited that the reconstructed $\rho(r)$ appears as a broad peak; the sharp edges characterising the square peak are lost. Thus X-ray waves scattered at high Q values carry high resolution information whereas waves scattered at lower Q values carry only general information about the rough shape of $\rho(r)$. Further discussion of this point occurs in Section V.4.

Another problem associated with resolution is illustrated in Fig. III.7. This figure shows that restricting the range of Q can produce spurious peaks in $\rho(r)$ reconstructed from $F(Q)$. In the figure peaks appear which were not present in Fig. III.4. These peaks have arisen solely because a limited range of $F(Q)$ values was used to calculate $\rho(r)$. The aim of an experiment is to find an unknown $\rho(\underline{r})$ from measurements of $F(\underline{Q})$ over a limited range of Q values. It is tempting to assign peaks, which might appear when equation III.1 is applied, to structural features in $\rho(\underline{r})$. This example shows that such a conclusion might be incorrect and that the peaks could simply arise from truncation errors.

In consequence refinement is best carried out as described in Section III.5 ie. in Q-space. In principle a trial model $\rho(\underline{r})$ can be used to apportion real and imaginary parts to $F(\underline{Q})$ whose modulus was measured experimentally. Thus the experiment could provide a source of amplitudes for the scattered waves and the trial model could produce phases. The experimental amplitudes and model phases could be used to compute a better approximation to $\rho(\underline{r})$. This is the basis of the heavy atom method of Section III.4 - where the trial model consists of the positions of the electron-dense atoms. Such a procedure can be misleading if only low resolution data are available because spurious peaks could be produced in $\rho(\underline{r})$. The approach described in Section III.5 is not subject to such errors since the model $\rho(\underline{r})$ is presumed to be defined at sufficient resolution for $F(\underline{Q})$ values to be calculated reliably in the range of Q values for which the diffraction pattern is to be analysed.

III.7. Loss of information at low Q values

Figure III.2 illustrates a general observation - that $I(\underline{Q})$ is very much higher at low Q values than elsewhere. A practical detector, such as a photographic film or a counter, has a limited dynamic range which then often precludes measurement of $I(\underline{Q})$ at low Q values if the rest of the pattern is to be measured accurately. Indeed $I(\underline{Q})$ is so great, for very low Q values, that, when it is detected by a photographic film, it exposes much of the surrounding area. X-Rays scattered at very low Q values are usually prevented from meeting the film by intercepting their path with a lead stop - which accounts for the white circles in the centres of the diffraction patterns in this book.

What is the effect of omitting $F(\underline{Q})$ corresponding to very low Q values from the integral in equation III.1? The same example is used as in Section III.6; Fig. III.4 represents the true structure and Fig. III.5 is its Fourier transform. Figure III.8 shows $\rho(r)$ reconstructed from $F(Q)$ omitting the $F(Q)$ values in question.

Two effects are apparent. Firstly, a roughly constant background, in this example about 0.2, is subtracted from $\rho(r)$. Thus the lowest reconstructed values of $\rho(r)$ are negative instead of being zero. Since in most experiments $F(\underline{Q})$ will not be measured on an absolute scale (see Section II.3), this scaling error is usually of little concern. Secondly, edges in the reconstructed $\rho(r)$ are enhanced

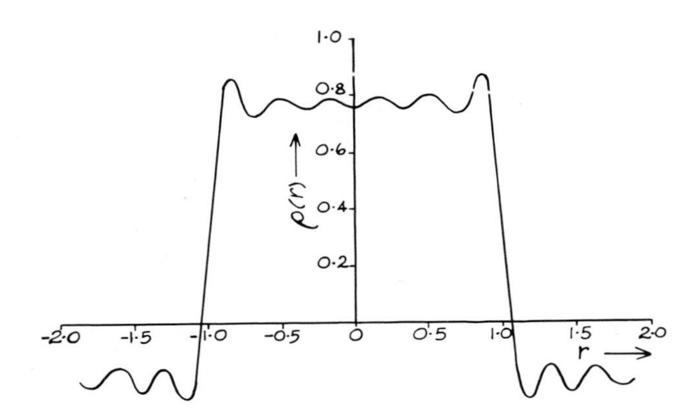

Fig. III.8. Square peak reconstructed from F(Q) of Fig. III.5 omitting Q in the range -0.2 to 0.2 reciprocal length units.

by a peak on one side and a trough on the other. It is important to note that this peak and trough are not real structural features, ie. they do not appear in Fig. III.4, but arise because a limited range of F(Q) values was used to compute ρ(r) in equation III.1.

III.8. Chirality

A chiral structure is one which cannot be superimposed on its mirror image; from the theory given so far it emerges that X-ray diffraction cannot generally distinguish a structure from its mirror image ie. the determination of chiral structures is ambiguous. It may not be immediately apparent that whether or not a structure is chiral can depend on the resolution at which it is being examined. Consider a protein unit which is made of four identical large molecules or sub-units. An experiment may be performed to determine whether these four sub-units are arranged at the corners of a tetrahedron or at the corners of a square - to make up a unit. Only low resolution data are required to answer this question ie. F($\underline{Q}$) may be confined to low Q values. If, at this low resolution, the sub-units are roughly spherical then the tetrahedral and square-planar structures are both achiral ie. they are identical to their mirror images. Thus the structure can be solved unambiguously. However each sub-unit consists of a polymer chain wound into some specific three-dimensional shape - rather like a somewhat irregular ball of wool. At this resolution the structure is chiral. If further diffraction data, corresponding to higher Q values, are added to the original set and used to determine the high resolution molecular structure of a sub-unit, they will be unable to distinguish the true structure from its mirror image.

The ambiguity which occurs when determining chiral structures arises because F($\underline{Q}$) and F*($\underline{Q}$) cannot be distinguished experimentally. From equation II.3 it follows that

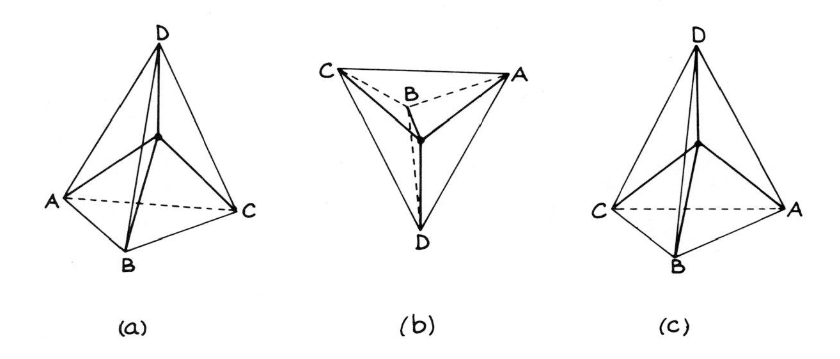

Fig. III.9. Inversion of a chiral structure.

$$F(-\underline{Q}) = \int \rho(\underline{r}) \exp(-i\,\underline{r}.\underline{Q})\, d\underline{r} \qquad \text{(III.7)}$$

$$= F^*(\underline{Q})$$

since $\rho(\underline{r})$ is real. The diffraction pattern, according to equation II.5, is a map of $F(\underline{Q})F^*(\underline{Q})$ which, according to equation III.7 is identical to $F(-\underline{Q})F^*(-\underline{Q})$ (Friedel's Law). Inspection of the diffraction pattern cannot then distinguish the "top" - which yields $F(\underline{Q})$ - from the "bottom" - which yields $F(-\underline{Q})$. Hence it is never clear, even if a structure is solved by deduction, whether one has detected $F(\underline{Q})$ or $F(-\underline{Q})$. Since $F(-\underline{Q})$ is identical to $F^*(\underline{Q})$, it is not known which of $F(\underline{Q})$ or $F^*(\underline{Q})$ has been determined; the former represents the correct set of phases but the latter represents the phases with their signs changed.

What happens if $F^*(\underline{Q})$ is used instead of $F(\underline{Q})$ in equation III.1? The result is

$$\int_S F^*(\underline{Q}) \exp(-i\,\underline{r}.\underline{Q})\, d\underline{Q}$$

$$= \int_{-\infty}^{\infty}\int_{-\infty}^{\infty}\int_{-\infty}^{\infty} F(-\underline{Q}) \exp(-i\,\underline{r}.\underline{Q})\, d\underline{Q} \qquad \text{[eqn.III.7]}$$

$$= -\int_{\infty}^{-\infty}\int_{\infty}^{-\infty}\int_{\infty}^{-\infty} F(\underline{Q}') \exp(i\,\underline{r}.\underline{Q}')\, d\underline{Q}' \qquad [\underline{Q}' = -\underline{Q}]$$

$$= \int_{-\infty}^{\infty}\int_{-\infty}^{\infty}\int_{-\infty}^{\infty} F(\underline{Q}') \exp(i\,\underline{r}.\underline{Q}')\, d\underline{Q}'$$

$$= \rho(-\underline{r}) \qquad \text{[eqn.III.1]}$$

Hence, if $F^*(\underline{Q})$ is used, equation III.1 returns the structure inverted in the origin.

If a chiral structure is inverted in a point the result is an upside-down mirror image. Figure III.9 shows a chiral molecule (a) in which A, B, C and D are at the corners of a tetrahedron. Inversion of this molecule in the central black atom yields (b). When (b) is turned the right way up and suitably rotated as in (c), it can be seen to be the mirror image of (a). Thus an experiment to determine an unknown chiral structure could yield not the correct result but its mirror image. In certain cases the ambiguity can be resolved by exploiting the properties of anomalous scattering (Sections V.5 and V.6).

III.9. Summary

Since the real and imaginary parts of $F(\underline{Q})$ cannot be measured separately, the inversion property of the Fourier transform cannot be used directly to calculate $\rho(\underline{r})$. This difficulty is termed the "phase problem" and has to be overcome by a process of deduction or by trial-and-error. Deductive methods are only applicable in certain favourable cases but the resulting models for $\rho(\underline{r})$ are established with greater certainty than in the trial-and-error approach. Both approaches can be aided by calculating the inverse Fourier transform of $I(\underline{Q})$ to yield the autocorrelation of $\rho(\underline{r})$.

X-Rays scattered at low Q values provide information about the rough shape of $\rho(\underline{r})$; high resolution information is carried by the X-rays scattered at higher Q values. When calculating $\rho(\underline{r})$ from low resolution data some of the peaks which appear may arise from truncation errors and not be true structural features. Omission of experimental data corresponding to very low Q values from such a calculation leads to subtraction of a roughly constant background from $\rho(\underline{r})$; edges are also enhanced by the appearance of peaks and troughs which are also not true structural features. X-Ray diffraction cannot normally distinguish a structure from its mirror image.

CHAPTER IV
Diffraction Geometry

IV.1. Introduction

In Chapter II it was established that scattered waves can be specified in both amplitude and phase by $F(\underline{Q})$. The direction of the vector $\underline{Q}$ is related to the scattering direction - its modulus, Q, is a function of the angle through which the wave is scattered. Also it was established that a detector measures $I(\underline{Q})$ which is the product of $F(\underline{Q})$ with its complex conjugate.

The practical problem which then arises is to relate the pattern of detected intensities to $I(\underline{Q})$. To rephrase the problem, each position on the detector has to be converted to a vector $\underline{Q}$ if the theory presented in Chapters II and III is to be applied to the analysis of the diffraction pattern. The purpose of this chapter is to show how positions on a detector correspond to positions in Q-space. Thus it serves to relate the theory of the previous chapters to a real diffraction pattern.

IV.2. Ewald sphere

An alternative interpretation of $\underline{Q}$ is implicit in Section II.3 which leads to the idea of an intensity distribution existing throughout an unlimited extent of a continuous Q-space - in much the same way as electron density is distributed in real space. To make the argument simpler, consider a one-dimensional Fourier transform which, from equation II.6, is given by

$$F(Q) = \int_{-\infty}^{\infty} \rho(r) \exp(i\, r\, Q)\, dr$$

We could represent $F(Q)$ by plotting a graph like Fig. III.5 of Section III.6. (Figure III.5 actually represents an example where $F(Q)$ is real and can be plotted on a single graph.) Considered in this way, Q is simply the abscissa against which the transform has to be plotted as ordinate - it can have any value between plus and minus infinity. $F(Q)$ "exists" in the space defined by this infinite extent of Q values in much the same way as $\rho(r)$ exists in a space defined by an infinite extent of r values.

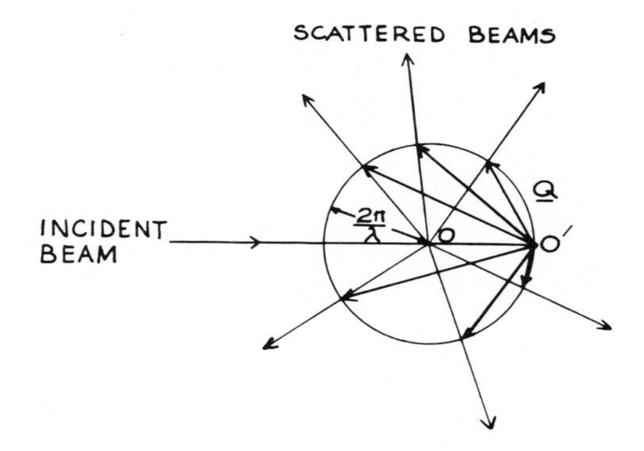

Fig. IV.1. Ewald sphere.

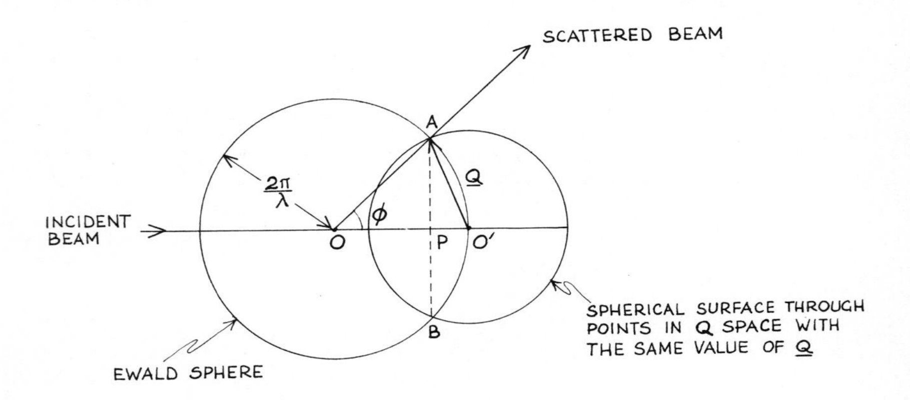

Fig. IV.2. Diffraction geometry for a spherically-averaged scatterer.

In three dimensions $\underline{Q}$ is simply the vector which defines a point at which we can compute the three-dimensional Fourier transform, $F(\underline{Q})$. The space defined by these vectors $\underline{Q}$ is called Q-space - we can define its origin (ie. the origin of the vectors) at the point O' of Fig. IV.1 - just as in Section II.3. From the relationship between $I(\underline{Q})$ and $F(\underline{Q})$, represented by equation II.5, we can consider that $I(\underline{Q})$ is capable of existing throughout the whole of this same Q-space. Thus there is a distribution of intensity which exists in Q-space, surrounding the origin O', much as there is a distribution of electrons surrounding the origin, O, of real space.

By measuring the intensity of scattered X-rays we can explore the distribution of intensity in Q-space. Since a diffraction experiment measures $I(\underline{Q})$, a diffraction pattern is a record of intensity distribution in a region of Q-space. But, as we shall see, not all of Q-space is accessible to a diffraction experiment.

Figure IV.1 shows an X-ray beam which is incident on a specimen whose origin is at O. In general the specimen will scatter X-rays in all directions. Each of these scattered directions corresponds to a vector $\underline{Q}$. According to Section II.3, the origin of all these vectors is at O' which is distant $2\pi/\lambda$ from O along the direction of the incident beam. Also according to this section, each of the vectors meets the scattered beam to which it corresponds at a distance $2\pi/\lambda$ away from O. Figure IV.1 then shows that, whatever the scattering direction, the vectors $\underline{Q}$ defined by scattered beams will describe a sphere of radius $2\pi/\lambda$ whose origin is at O - the Ewald sphere.

In consequence a diffraction experiment only provides values of $I(\underline{Q})$ at those points in Q-space which lie on the surface of the Ewald sphere - taking the incident beam to be fixed. The reason is that scattered X-rays can only correspond to those $\underline{Q}$ vectors which terminate at the surface of this sphere. We can therefore consider a diffraction pattern as being formed by the intersection of the Ewald sphere with the distribution of intensity in Q-space; this model explains the appearance of diffraction patterns from gases, liquids and amorphous solids. It will be seen, in Chapters VI and VII, that, for these systems, $\rho(\underline{r})$ has spherical symmetry albeit of a statistical nature. Consequently $I(\underline{Q})$ depends only on the modulus of $\underline{Q}$, not on its direction, and can be written as $I(Q)$. Figure IV.2 shows a spherical shell in Q-space; if $I(Q)$ is spherically symmetric, it will have the same value all over this shell. In the figure the shell intersects the Ewald sphere at A and B; in three dimensions A and B will lie on a circle, whose centre is at P, which will have a uniform intensity. To make this description clearer - the circle encloses a plane surface which is perpendicular to the plane of Fig. IV.2. Thus, for a scatterer with spherical symmetry, the diffraction pattern has circular symmetry. Figure IV.3 shows an example of such a pattern; the distribution of intensity along a radius of this pattern is analogous to a graph of $I(Q)$ plotted against Q as, for example, in Fig. III.2.

IV.3. Q-space

Figure IV.4 shows an X-ray beam incident on a specimen. Much of this beam continues, in a straight line, to a point C on a plane detector (usually a photographic film) which is perpendicular to the incident beam direction. Some radiation will be scattered so that it meets the detector at other points. Consider a scattered beam which meets the detector at the point E; this point may be defined by the vector $\underline{x}$.

The problem is to relate the vector $\underline{x}$ on the detector to the vector $\underline{Q}$ which specifies the scattered beam. Figure IV.4 shows that $\underline{x}$ is in the same plane as the incident and scattered beams as, according to its definition, is $\underline{Q}$. Thus the

Fig. IV.3. X-Ray diffraction pattern of paraffin wax.

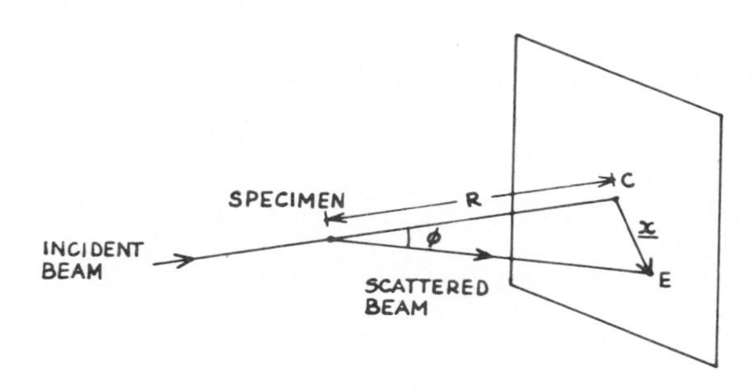

Fig. IV.4. Recording a diffraction pattern on a flat film perpendicular to the incident X-ray beam.

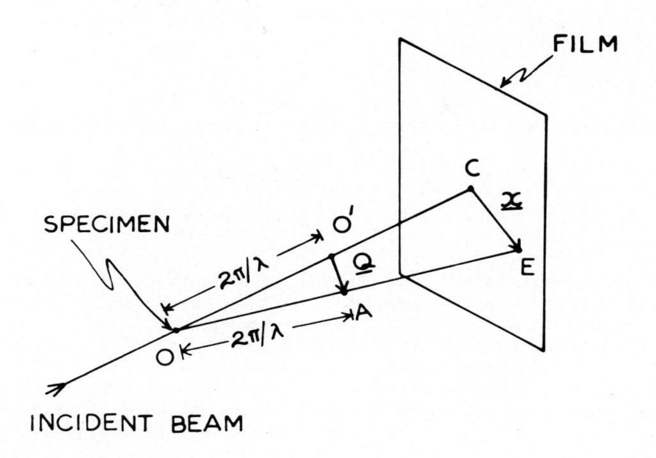

Fig. IV.5. Relationship between a point on the film and a point in Q-space.

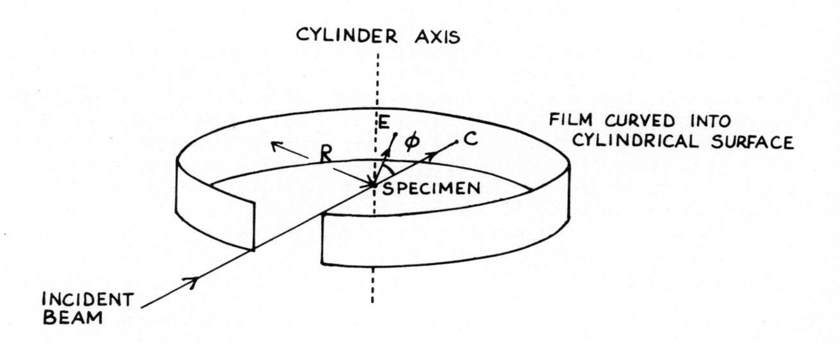

Fig. IV.6. Recording a diffraction pattern from a spherically symmetric specimen on a cylindrical film.

direction of $\underline{x}$ on the film indicates the direction of the vector $\underline{Q}$ from O' to some point on the Ewald sphere in Fig. IV.1. More rigorously, $\underline{x}$ lies in the same direction as the projection of $\underline{Q}$ on to the detector - see Fig. IV.5.

The modulus of $\underline{Q}$ can be calculated from x, the modulus of $\underline{x}$. If R is the distance from the specimen to the film, then

$$\phi = \tan^{-1} (x/R) \qquad \text{(IV.1)}$$

Equation IV.1 follows from Fig. IV.4. From equations II.4 and IV.1

$$Q = (4\pi/\lambda) \sin [(1/2) \tan^{-1} (x/R)] \qquad \text{(IV.2)}$$

Thus both the direction and the modulus of $\underline{Q}$ can be deduced from the diffraction pattern.

Figure IV.6 shows a common arrangement for measuring the diffraction pattern from a specimen with spherical symmetry. Such a pattern has circular symmetry, as shown in the example of Fig. IV.3, for the reasons given in Section IV.2. The distribution of intensity along any radius of the diffraction pattern then contains all the required information concerning the dependence of I(Q) on Q. A diameter of the pattern can be recorded on a strip of photographic film. If the film is bent into a cylinder with the specimen on its axis, as shown in Fig. IV.6, it is trivial to calculate ϕ from the radius, R, and the arc length, CE; Q can then be calculated from ϕ using equation II.4. A counter is often used instead of a photographic film with this arrangement. The counter rotates in a circular arc centred on the specimen and thus measures intensity as a function of angle. Equations which form the basis for calculating Q from a variety of other detector geometries are given in Volume II of the "International Tables for X-Ray Crystallography" (see BIBLIOGRAPHY Section 2).

What units are used to measure distances in Q-space? Three different systems are used. The one adopted in this book is commonly used in solid- and liquid-state physics where $\underline{Q}$ is sometimes denoted by $\underline{K}$. According to equation II.4, Q has the units of reciprocal length (usually nm^{-1}). Crystallographers tend to use a vector $\underline{S}$ which is related to $\underline{Q}$ by

$$\underline{S} = \underline{Q}/2\pi$$

S also has the dimensions of reciprocal length but it makes the interpretation of diffraction patterns from periodic structures simpler. In Section VIII.3 it will be shown that the diffraction pattern from a series of identical scatterers spaced a distance c apart is confined to lines spaced $2\pi/c$ apart in Q-space. Now if the pattern is described by the vector $\underline{S}$, these lines are $1/c$ apart in S-space. Thus distances in S-space can be immediately inverted to yield repeat distances in regularly periodic structures. This simplifying feature leads to its adoption by crystallographers who refer to S-space as "reciprocal space". A few crystallographic texts make the diffraction space dimensionless by defining

$$\underline{\Omega} = \lambda\underline{Q}/2\pi$$

Ω-space is also referred to as reciprocal space and is measured in the dimensionless "reciprocal lattice units". When using formulae which relate distances in Q-space to spacings in a scatterer, it is essential to convert measurements from the diffraction pattern into the correct space ie. the one which the formula applies to. In this book all formulae are developed in Q-space (as opposed to S-space or Ω-space).

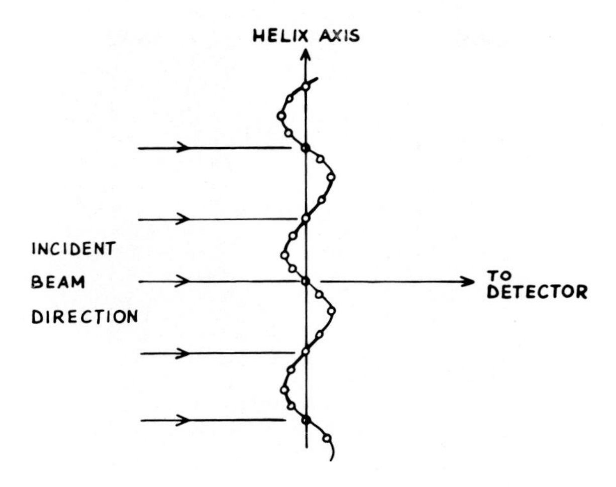

Fig. IV.7. X-Ray beam incident at right angles to the axis of a helix of atoms.

IV.4. Simulation by optical diffraction

Diffraction of light by an array of apertures can be used to simulate the diffraction of X-rays by an array of atoms or molecules. This simulation is particularly useful in those cases where $F(\underline{Q})$, and hence $I(\underline{Q})$, would be difficult to compute. Using Huygen's construction, an aperture can be considered as a source of secondary waves in the same way that electrons emit secondary X-rays when they are excited by an incident beam.

The region of interest in the optical simulation is confined to low Q values. Although λ for X-rays is of the order of distances between atoms, the dimensions of the apertures used for optical diffraction are conventionally about 10^4 times greater than λ for light. Thus optical diffraction data can provide higher resolution information about the arrangement of these apertures than X-ray diffraction can provide about an arrangement of atoms. Section III.6 provides the detailed justification for this statement - the maximum value of Q which can be measured, and which therefore determines the ultimate resolution of a diffraction experiment, is inversely proportional to λ. If we are to compare our optical diffraction analogue with an X-ray diffraction pattern we need only consider the lower Q values which are theoretically accessible to light.

In the optical case the array of apertures is two-dimensional whereas in the X-ray case the array of atoms is three-dimensional; this difference does not invalidate the simulation. We need only consider very low Q values in the optical case. The area of the Ewald sphere for light which gives rise to this region of the diffraction pattern is so small that it is effectively planar. Thus the optical diffraction pattern can be considered to be formed by a plane intersecting $I(\underline{Q})$; this plane passes through O', a distance $2\pi/\lambda$ from the origin of real space, and is

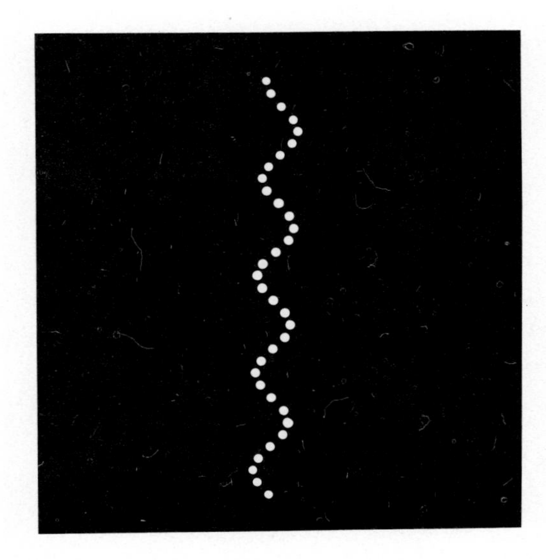

Fig. IV.8. Apertures punched in an opaque screen to represent the arrangement of atoms in the helix of Fig. IV.7 projected on to a plane perpendicular to the incident beam direction.

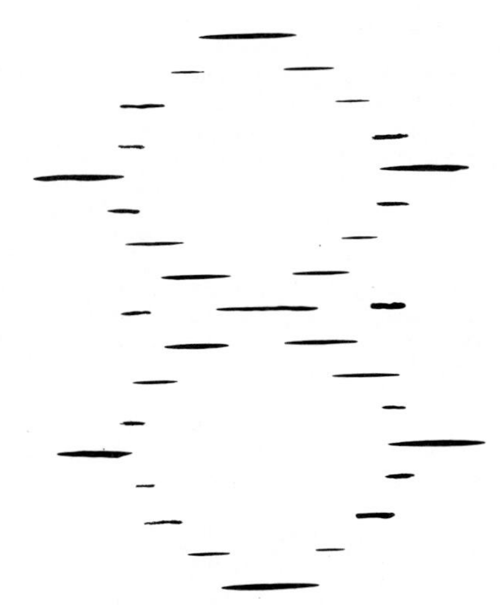

Fig. IV.9. Optical diffraction pattern from the array of apertures in Fig. IV.8.

normal to the incident beam - from Section IV.2. A property of the Fourier transform is that a plane section through $F(\underline{Q})$, which contains the origin of Q-space, is equivalent to the transform of $\rho(\underline{r})$ projected on to a parallel plane in real space. Therefore the optical diffraction pattern from a three-dimensional scatterer is exactly the same as that from the projection of its structure on to a plane perpendicular to the incident direction. The complication arising in the diffraction of light from large bodies, discussed in Section II.5, does not arise here because the medium of propagation is unchanged by passage through an aperture in an opaque screen.

How is the technique used in practice? Suppose an X-ray diffraction pattern which has been recorded from a specimen has to be solved by trial-and-error, as in Section III.5. A trial model has also been proposed - that the structure is a helix of atoms whose axis is perpendicular to the incident X-ray beam, as in Fig. IV.7. What does the X-ray diffraction pattern of this trial model look like? We could build an actual scale model, using suitable spheres to represent atoms, and examine the way it scattered a beam of light incident at right angles to the helix axis. Apart from the theoretical problems raised in Section II.5, this experiment would be extremely tedious to perform. All we need do is to punch a set of apertures which represent the positions of the atoms in the structure projected on to a plane perpendicular to the incident beam direction - as in Fig. IV.8. The optical diffraction pattern of this "mask", in Fig. IV.9, has the appearance we would expect from the X-ray diffraction pattern of our trial model - if the model is acceptable. Optical diffraction can also be used to investigate the properties of Fourier transforms in two dimensions in the same way and is useful in cases where calculation might be unduly complicated.

IV.5. Summary

An X-ray diffraction pattern can be considered to be formed when a sphere of radius $2\pi/\lambda$, centred at the origin of real space, intersects Q-space. This sphere is called the Ewald sphere. The pattern is recorded by projecting this intersection on to the detector. For a specimen with spherical symmetry, like a liquid or a gas, the pattern will therefore have circular symmetry and be centred about the point where the undeflected beam meets the detector. The direction and magnitude of the vector $\underline{Q}$ can be measured from the geometry of the detector system. Diffraction of light by a two-dimensional mask can be used to investigate the properties of Fourier transforms and to simulate X-ray diffraction.

CHAPTER V

Atoms and Molecules

V.1. Introduction

The wavelengths of X-rays are of the order of the distances between atoms in molecules eg. copper K_α X-rays have λ = 154 pm and the carbon-to-carbon single bond is also around 154 pm long. Thus, if data are collected to sufficiently high Q values, an X-ray diffraction experiment can provide information on the arrangement of atoms in molecules. X-Ray diffraction data which are restricted to lower Q values provide information on the structure of the scatterer at lower resolution (Section III.6). Hence these "low angle" data are sensitive to the gross shapes of molecules and to their arrangements in scattering specimens.

In consequence X-ray diffraction is used to investigate the structure of matter at the molecular level. Such investigations include the arrangement of atoms in molecules (ie. molecular structure), the overall shapes of molecules (eg. whether a particular macromolecule is rod-shaped or globular) and the way molecules are arranged in systems. In order to pursue these investigations it is necessary to understand how atoms and molecules scatter X-rays.

V.2. Atoms

Suppose that the electron density in an atom has spherical symmetry. This symmetry can be considered as a time average over the course of an X-ray diffraction experiment. Its Fourier transform will, according to Section II.4, also have spherical symmetry and is consequently a function of the modulus of $\underline{Q}$ but not its direction. This Fourier transform is usually denoted by $f(Q)$ and is called the "atomic scattering factor". According to equation II.4 it is given by

$$f(Q) = \int_0^\infty 4\pi r^2 \rho(r) \operatorname{sinc}(Qr)\, dr \qquad (V.1)$$

where there is considered to be only the one atom in accessible space.

In principle $\rho(r)$ can be calculated for an atom of each element; equation V.1 can then be used to calculate $f(Q)$. Calculation of $\rho(r)$ is clearly a quantum mechanical problem but equation V.1 was derived by ignoring complications arising from the quantum theory of scattering - these complications are discussed briefly

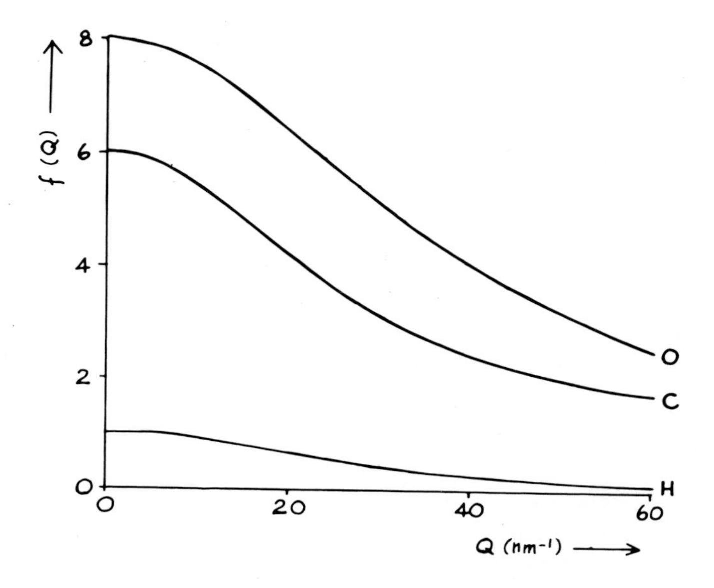

Fig. V.1. Atomic scattering factors for oxygen, carbon and hydrogen

in Section I.5. Application of the quantum theory at this late stage in the calculation of f(Q) is not, therefore, really justifiable; a thorough account of the rigorous calculation is given in the book by James (see BIBLIOGRAPHY Section 2). It is perhaps worth noting that the classical theory used here proves useful only because the scattering electrons in matter are bound in atoms. Volume IV of the "International Tables for X-Ray Crystallography" (see BIBLIOGRAPHY Section 2) lists atomic scattering factors and also gives estimates of their reliability. For aluminium the experimental and theoretical values of f(Q) differ by several percent; differences of this kind could introduce small systematic errors into the determination of molecular structures. Figure V.1 shows a few examples of the dependence of atomic scattering factors on Q.

Fortunately it is possible to deduce most of the properties of the atomic scattering factor simply by inspecting the form of equation V.1. Since the atom is considered to have spherical symmetry, f(Q) is real. Also it is always positive which means that, according to Section II.4, all atoms scatter X-rays with the same phase for all Q values. In Section V.5 a complication will be described but it usually only affects the analysis of X-ray diffraction patterns when a chiral structure is to be distinguished from its mirror image.

As might be expected, the greater the number of electrons in the atom - the greater is the scattering factor; in particular when Q equals zero the scattering factor is equal to the number of electrons in the atom. The number of electrons at a distance r from the centre of an atom is given by

$$U(r) = 4\pi r^2 \rho(r) \tag{V.2}$$

From equations V.1 and V.2 the atomic scattering factor becomes

$$f(Q) = \int_0^\infty U(r)\ \text{sinc}\ (Qr)\ dr \tag{V.3}$$

In the limit as x tends to zero, sinc(x) tends to unity - so that

$$f(0) = \int_0^\infty U(r)\ dr = Z$$

where Z is the number of electrons in the atom. Thus the amplitude scattered with a zero value of Q by an atom is equal to its number of electrons - given the definition of F(Q) in equation II.3 which neglects some fundamental constants.

The fall-off of f(Q) with increasing Q is caused by interference between X-rays scattered in the same direction by different parts of the same atom. In contrast the atomic scattering factor for thermal neutrons does not decrease appreciably as Q increases. The reason is simply that neutrons, unlike X-rays, are scattered by the atomic nucleus which is very small compared with λ; interference effects between neutrons scattered by the same atom are therefore negligible.

V.3. Molecules

The term "molecule" here simply implies an assembly of atoms in close association which it is convenient to consider together for the purpose of calculating $F(\underline{Q})$. Thus a molecule in this context does not necessarily imply a covalently bonded structure. In Chapters VIII and X the theory developed here for a molecule will be applied to the entire structure which repeats itself regularly to form a crystal. Elsewhere in this book, however, the theory will usually be applied to molecules in the chemical sense ie. a covalently bonded assembly of atoms.

When an assembly of atoms scatters X-rays, interference effects will occur between those X-rays scattered in the same direction by different atoms. This inter-atomic interference is in addition to the intra-atomic interference effects which lead to the fall-off in f(Q) with increasing Q values. According to Section II.4 the Fourier transform of a molecule is simply obtained by adding together the transforms of its constituent atoms with due regard to the phase differences introduced by the positions of the atoms. Then, for a molecule containing N atoms, the molecular transform is given by

$$F(\underline{Q}) = \sum_{j=1}^{N} f_j(Q)\ \exp\ (i\ \underline{r}_j \cdot \underline{Q}) \tag{V.4}$$

where $\underline{r}_j$ is the position of the j th atom, related to some arbitrarily chosen origin, and f_j is its atomic scattering factor.

The complex $F(\underline{Q})$ in equation V.4 represents the amplitude and phase of the X-rays scattered, with a particular vector $\underline{Q}$, by a molecule. Its form allows for interference effects between X-rays scattered by different atoms while f, which depends only on the modulus of $\underline{Q}$, allows for interference effects between X-rays scattered by different parts of the same atom. Note that the phase, but not the amplitude, of $F(\underline{Q})$ depends on where the origin is chosen - this property is physically reasonable since definition of phase, but not of amplitude, requires some arbitrarily chosen reference. When Q equals zero, $F(\underline{Q})$ is simply the sum of all the atomic scattering factors which, according to Section V.2 is equal to the number of electrons in the molecule.

It follows, from equation V.4, that a chiral molecule cannot be distinguished from its mirror image since $f(Q)$ is real and positive. In Section III.8 it was shown that $F(\underline{Q})$ could not be distinguished experimentally from $F(-\underline{Q})$ - because the number of electrons in a unit volume of space, ie. $\rho(\underline{r})$, must always be positive. Consequently a chiral structure could not be distinguished from its mirror image. Exactly the same reasoning applies here - thus the expression for $F(\underline{Q})$ of a molecule in equation V.4 has the general property described previously. Section V.5 describes circumstances under which the simple approach used so far breaks down and the chirality of a molecule can then be determined using X-ray diffraction data from a suitable system.

V.4. Molecules at low resolution

According to Section III.6, $F(\underline{Q})$ will be insensitive to detailed molecular structure at sufficiently low Q values. To rephrase this statement - waves scattered at these low values of Q correspond to a low resolution view of the structure. In order to explain these "low-angle" diffraction data, it is often sufficient to consider the molecule to have some simple shape with a uniform electron density ie. to ignore those details of the structure which cause variations of $\rho(\underline{r})$ within the molecule. Thus some globular macromolecules, such as certain enzymes, may be considered as spheres and certain polymers simply as rods in the analysis of low-angle X-ray diffraction patterns.

X-Ray scattering by spheres of uniform electron density is particularly applicable as will be seen in Section VI.3. From Section II.4 it follows that the Fourier transform of the sphere depends only on the modulus of $\underline{Q}$ and is independent of its direction. Then, according to equation II.8

$$F(Q) = \int_0^a 4\pi r^2 \rho \ \mathrm{sinc}\ (Qr)\ dr$$

for a sphere of radius a with a uniform electron density, ρ. This integral evaluates to

$$\left.\begin{aligned} F(Q) &= (4\pi a^3 \rho/3)\ \Phi(Qa) \\ \Phi(x) &= 3(\sin x - x \cos x)/x^3 \end{aligned}\right\} \qquad (V.5)$$

Equation V.5 may be simplified to

$$\left.\begin{aligned} F(Q) &= n\ \Phi(Qa) \\ n &= 4\pi a^3 \rho/3 \end{aligned}\right\} \qquad (V.6)$$

noting that n, in equation V.6, is the number of electrons in the sphere.

Equation V.6 shows the same kind of behaviour as the exact expression for the Fourier transform of a molecule at low Q values. Figure V.2 shows the form of the function $\Phi(x)$. Note that, as x tends to zero, $\Phi(x)$ tends to unity. Therefore, when Q equals zero, $F(Q)$, in equation V.6, is simply equal to the number of electrons. Thus this approximate form for the molecular transform yields the same result as the exact expression of equation V.4 in the limiting case of Q equals zero. The dependence of $F(Q)$ for a sphere on Q, which arises mathematically from the form of Φ in Fig. V.2, can be considered physically as the result of interference effects between X-rays scattered in the same direction by different parts of the sphere. As this simple model for the molecular transform ignores the fact

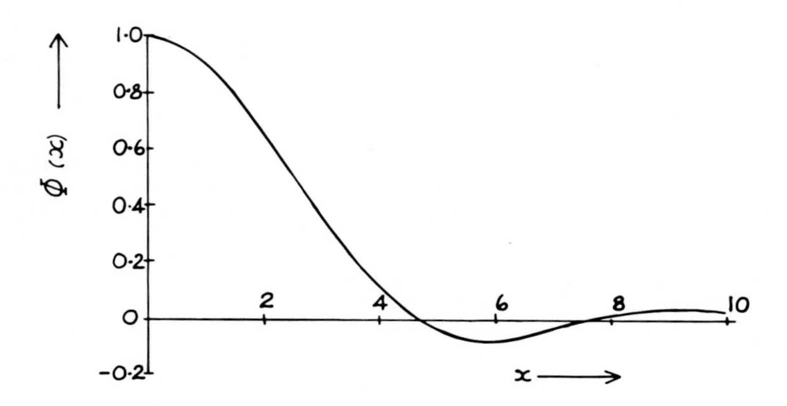

Fig. V.2. The function Φ(x) defined in equation V.5.

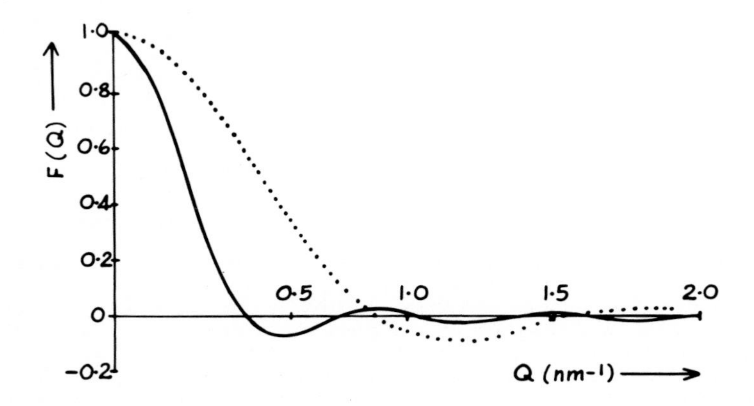

Fig. V.3. Comparison of F(Q) for spheres of radius 2.5 nm (dotted line) and 10 nm (continuous line) computed from equation V.6 with n equal to unity.

that molecules consist of atoms, it becomes a poorer approximation for higher resolution diffraction data ie. at higher Q values. Figure VI.1 of Section VI.3 provides an example which demonstrates the application of equation V.6 as an approximation to the molecular transform and shows how its validity depends on Q.

Three further properties of the Fourier transform of a sphere can be simply deduced from equation V.6. Firstly, although F(Q) in equation V.6 is not complex it still conveys phase information. In Fig. V.2, $\Phi(x)$ crosses the x axis periodically. Thus F(Q) must cross the Q axis - corresponding to a change in phase of the scattered X-rays as explained in Section II.4. A change from a positive to a negative value of F(Q) corresponds to a phase change from 0 to π radians and vice versa. Secondly, since $\Phi(x)$ is equal to $\Phi(-x)$, F(Q) in equation V.6 is the same as F(-Q).

The third property of F(Q) in equation V.6 provides an example of a general property of Fourier transforms. Figure V.3, which was computed using equation V.6, shows that F(Q) varies more for a large sphere than for a smaller sphere. This observation is a special case of a general property - that large dimensions in real space give rise to rapidly varying changes in F(Q). Conversely, gentle variations in F(Q) correspond to small distances in real space. It is this property which leads to the observation of Section III.6 - that $F(\underline{Q})$ does not carry high resolution information at low Q values. Gentle variations in $F(\underline{Q})$ can only be detected if a wide range of Q values is examined; if the examination of Q-space is restricted to low Q values these variations will be undetected and high resolution information lost.

V.5. Anomalous scattering

This section is concerned with a phenomenon which may be used to distinguish a chiral structure from its mirror image in favourable circumstances; in the course of explaining this phenomenon some of the statements of Sections I.3 and II.2 will be justified. Both this section and Section V.6 may be omitted by readers who are not interested in determining the chirality of structures; readers who are interested in this topic, but who dislike mathematical explanations, can proceed directly to Section V.6. Explanations in this section rely, as far as possible, on classical theory for the reasons given in Section I.5; the quantum theory of anomalous scattering is described, and compared with classical theory, in the book by James (see BIBLIOGRAPHY Section 2).

In Fig. I.1 of Section I.3 the electron oscillates up and down the z-axis as it is accelerated by the oscillating electric field of the incident radiation. Since the electric vector undergoes simple harmonic oscillation, we might also expect the electron to be a simple harmonic oscillator. The equation of motion for an undamped, undriven electron oscillating along the z-axis in this fashion is simply

$$d^2z/dt^2 = - \omega_o^2 z \tag{V.7}$$

where ω_o is its natural angular frequency ie. the value in the absence of damping or driving forces. However the motion of an electron in an atom will be damped eg. by the attractive force exerted by the nucleus. In many similar phenomena this damping is proportional to the velocity of the oscillator; thus equation V.7 has to be modified to

$$d^2z/dt^2 + k\, dz/dt + \omega_o^2 z = 0 \tag{V.8}$$

where k is a constant for a particular electron - the "damping factor". Now the electron is being constantly driven by the oscillating field of the incident X-rays. Since we are considering oscillations which are confined to the z-axis, the field at some instant in time, t, can be denoted by a scalar, E, whose oscillatory nature can be shown by writing it in the form

$$E = E_o \exp (i \omega t)$$

where ω is the angular frequency and E_o the amplitude of the oscillation. The force which a field E exerts on the charge, e, of the electron is simply eE. If m is the mass of an electron, its acceleration is then

$$eE/m = (e/m) E_o \exp (i \omega t)$$

Therefore equation V.8 has to be further modified to

$$d^2z/dt^2 + k\, dz/dt + \omega_o^2 z = (e/m) E_o \exp (i \omega t) \qquad (V.9)$$

Equation V.9 can be solved for the displacement, z, of the electron. The result

$$z = (e/m) E_o \exp (i \omega t)/(\omega_o^2 - \omega^2 + i k \omega) \qquad (V.10)$$

can easily be verified by substitution into equation V.9.

Equation V.10 will be used to derive the amplitude of the X-rays emitted by the oscillating electron. According to classical electrodynamics, the electric field of the radiation emitted by an electron with an acceleration of d^2z/dt^2 is given by

$$E_s = (\mu_o/4\pi)(e/R)(\sin \phi_z)\, d^2z/dt^2 \qquad (V.11)$$

where μ_o, R and ϕ_z are defined in Section I.3. Differentiating equation V.10 twice with respect to time and substituting the result into equation V.11 yields the electric field of the emitted electromagnetic wave

$$\left.\begin{aligned} E_s &= - C\omega^2/(\omega_o^2 - \omega^2 + i k \omega) \\ C &= E(\mu_o/4\pi)(e^2/mR) \sin \phi_z \end{aligned}\right\} \qquad (V.12)$$

The intensity of the X-rays scattered by the electron can be calculated from equation V.12. Following Section II.3 the result is

$$\left.\begin{aligned} I_s &= \Omega I_o (\mu_o/4\pi)^2 (e^4/m^2R^2) \sin^2 \phi_z \\ \Omega &= \omega^4/[(\omega_o^2 - \omega^2)^2 + k^2\omega^2] \end{aligned}\right\} \qquad (V.13)$$

where I_o is the incident intensity; in this approach E_s is taken to represent a wave whose intensity is given by $E_sE_s^*$. (Unfortunately this approach ignores the contribution of the magnetic vector to the intensity. Because of the relationship between the electric and magnetic vectors a constant of proportionality then arises in obtaining I_s from E_s and I_o from E. Of course this constant appears on both sides of equation V.13 and hence cancels out. Details are given in the book by Bleaney and Bleaney - pp. 260 to 265 - see BIBLIOGRAPHY Section 7). Equation V.13 is identical to equation I.1 - apart from the appearance of Ω. Provided ω is greater than k, Fig. V.4 shows that Ω is very close to unity when ω is greater than ω_o. Thus equation I.1 gives I_s provided that the angular frequency of the

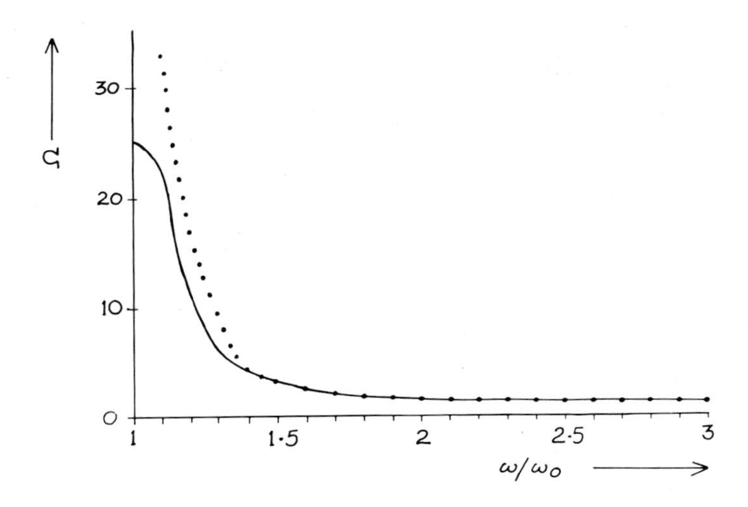

Fig. V.4. Ω plotted against ω/ω_o for ω_o/k equal to 5 (continuous line) and 50 (dotted line).

incident X-rays is greater than the natural angular frequency of the electron in its atom. As ω approaches ω_o, I_s will suddenly increase - because of the behaviour of Ω shown in Fig. V.4. Thus ω_o can be considered as an angular resonant frequency.

This resonance effect can be used to explain the discontinuities in the absorption coefficient of an element (its absorption edges) described in Section I.2. The absorption coefficient of an element is defined empirically. If a monochromatic X-ray beam of intensity I_o is incident on a thickness, T, of material, the emergent intensity is given by

$$I = I_o \exp(-\mu T) \qquad (V.14)$$

where μ is a constant for a given material - its "absorption coefficient". A material which has a high value of μ absorbs a high proportion of the incident X-rays. According to Section I.5 we can consider an X-ray beam as a stream of photons of energy hc/λ where h is Planck's constant, c is the speed of electromagnetic radiation in a vacuum and λ is the wavelength of the X-rays in the wave description. Thus the greater the value of λ, the lower is the energy of the X-ray photons and, consequently, the less easily they would be expected to pass through matter. Figure V.5 shows that, as expected, μ increases with λ - except that its value suddenly drops when λ equals a critical value, λ_o. In Fig. V.6 this behaviour of μ is combined with Fig. I.1 to show how a nickel filter can be used to remove copper K_β X-rays.

How is the sudden decrease of μ in Fig. V.5 explained by resonance? The transmitted intensity can be attributed to scatter for which Q equals zero and obtained from equation V.13 by giving ϕ_z the value of 90° ie. sin ϕ_z becomes unity. For a given incident intensity, I_o, the transmitted intensity then increases when Ω increases ie., for a high value of ω, when ω decreases to approach ω_o. If the transmitted intensity increases, the absorption coefficient, μ, will decrease.

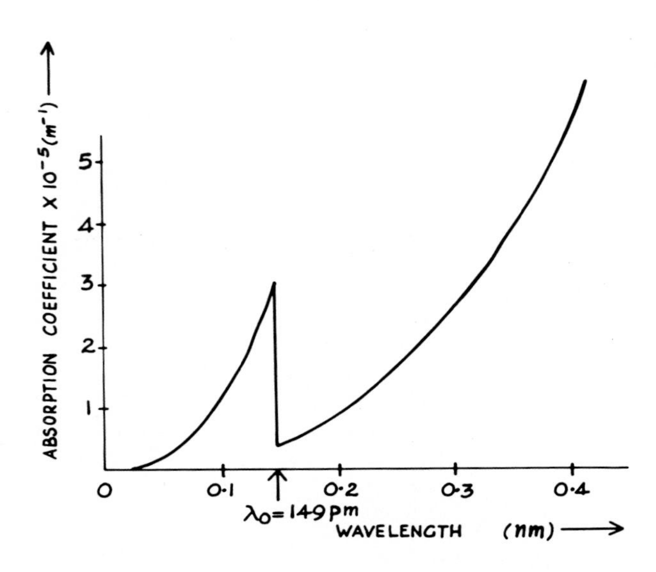

Fig. V.5. Absorption coefficient for nickel as a function of wavelength.

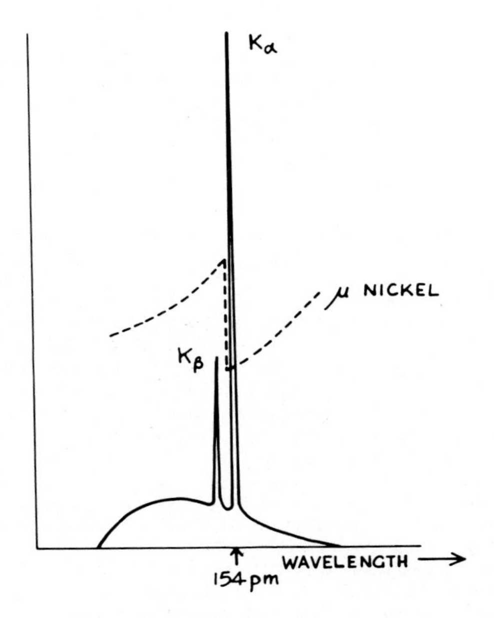

Fig. V.6. Filtering copper K_{α} X-rays with nickel.

How can the decrease in μ, when ω approaches ω_o, be related to λ? Since ω and ω_o are angular frequencies, they are related to frequency, ν, and hence to λ by

$$\omega = 2\pi\nu = 2\pi c/\lambda$$

Thus a wavelength λ_o is associated with ω_o and hence with the decrease in μ - λ_o can be used to define the position of the sudden decrease, or absorption edge, which appears in Fig. V.5. Resonance occurs when λ is such that the energy of an incident X-ray photon equals the binding energy of an electron in an atom; the electron is then emitted, following collision with a photon, and another falls from a higher energy level to take its place. (When the photon energy exceeds the binding energy such an event becomes less probable and resonance does not occur.) Thus each energy level has its corresponding value of λ_o. For elements whose atomic number is below about 50, the so-called "K-edge" is the closest to the X-ray wavelengths used for diffraction experiments. Also λ_o, for the K-edge, is greater than λ of copper K_α X-rays (ie. ω is greater than ω_o) for elements up to atomic number 27 (cobalt).

If the wavelength, λ, of the X-rays incident on an atom is close to λ_o, the position of its absorption edge, the atomic scattering factor becomes complex. The atomic scattering factor was shown to be real in Section V.2. In fact this result arose from the assumption, in Section II.2, that the amplitude of the X-rays scattered by an electron in matter was real ie. it was assumed that the amplitude, E_s, in equation V.12 was real. By taking the positive square root it was inevitable that f(Q) in Section V.2 would be real and positive. In Section II.2 the immediate result of taking the positive square root was that $F(\underline{Q})$ would always be real and positive when Q equalled zero ie. the X-rays scattered with Q equal to zero were assigned a phase of zero radians; this scattered beam became the reference for all phase measurements and, according to Section I.3, the phase of the incident beam became $-\pi$ radians. But, according to equation V.12, E_s is complex.

When λ_o is greater than λ, f(Q) is real and positive and is given by equation V.3. Figure V.4 shows that Ω tends to unity when ω is greater than ω_o ie. λ_o is greater than λ. Also the theory used to explain absorption edges depends on ω being greater than k. When ω is much greater than ω_o (which is the case for atoms of low atomic number), the only way Ω, as defined in equation V.13, can then approach unity is for ω_o and $k\omega$ to become effectively zero. Then E_s in equation V.12 will be real and reduces to the positive square root of I_s in equation I.1. According to the previous paragraph, f(Q) is then given by equation V.3. When ω_o is much greater than ω, ω_o^2 will be very much greater than k and, once again, the imaginary part of f(Q) in equation V.14 disappears. In this case E_s for the electrons in question will be negative - since ω_o^2 in equation V.12 is much greater than ω^2. However, ω_o will be much greater than ω only for atoms of high atomic number. The scatter from all the other electrons will then predominate and for most of them ω_o will be much less than ω. Consequently f(Q) is still real and positive.

If λ is close to λ_o, f(Q) is complex and equation V.3 has to be modified. Figure V.4 shows that Ω is very much greater than unity when ω is close to ω_o ie. λ is close to λ_o - thus k cannot be neglected in equation V.12. Since Ω is large and E_s is complex, the contribution of the resonating electrons to f(Q) is complex. Also the magnitude of the real part of E_s will be affected. It is therefore customary to correct f(Q) for anomalous scattering by writing

$$f_c(Q) = f(Q) + f'(Q) + i\, f''(Q) \qquad (V.15)$$

Values of f' and f" are given in Volume III of the "International Tables for X-Ray Crystallography" (see BIBLIOGRAPHY Section 2) but for many purposes they are negligible.

V.6. Determination of chirality

If f(Q) is real, ie. when the wavelength of the incident X-rays is far from an absorption edge of the atom, a chiral molecule cannot be distinguished from its mirror image ie. its chirality cannot be determined. Suppose a structure is determined by a method of the kind described in Chapter III. In Section III.8 it was shown that it is not then clear whether the resulting electron density, $\rho(\underline{r})$, or its inverse, $\rho(-\underline{r})$, represents the true structure. One yields the true structure of a molecule: the other gives its mirror image. The reason for the ambiguity was shown to be that, since $\rho(\underline{r})$ is real, it is impossible to distinguish $I(\underline{Q})$ from $I(-\underline{Q})$ experimentally. If all the atomic scattering factors, $f_j(Q)$, in equation V.4 are real, the arguments of Section III.8 still apply.

The chirality of a molecule can be determined when one, or more, of the atomic scattering factors is complex ie. when it contains an atom with an absorption edge which is close to the wavelength of the incident X-rays. Then the arguments of Section III.8 do not apply to $F(\underline{Q})$, as defined by equation V.4, and there is no longer any condition for $I(\underline{Q})$ to equal $I(-\underline{Q})$.

How is the chirality of a molecule determined in practice? A structure determination yields two solutions - $\rho(\underline{r})$ and $\rho(-\underline{r})$. A set of $F(\underline{Q})$ values is then calculated from each solution, for positive and negative values of Q, using equation V.4; these calculations employ the complex atomic scattering factors of equation V.15. Each set of $F(\underline{Q})$ values is used to calculate a set of $I(\underline{Q})$ values using equation II.5 - once again for both positive and negative values of Q. The set of calculated $I(\underline{Q})$ values which agrees better with those measured experimentally must have been calculated from the better model for the structure.

The application of anomalous scattering to the determination of chirality is usually restricted to X-ray diffraction data from crystals. The reason is that $I(\underline{Q})$ and $I(-\underline{Q})$ must not overlap on the diffraction pattern - otherwise they cannot be separated for comparison with calculated values. Because of the arrangements of molecules in most phases of matter, $I(\underline{Q})$ and $I(-\underline{Q})$ will overlap; but they can be separated experimentally when recording diffraction patterns from crystals.

V.7. Summary

The resultant X-ray wave scattered by a given type of atom depends only on the modulus of the vector $\underline{Q}$. All atoms scatter a resultant wave with the same phase which does not vary with Q. For a given atom type the wave is represented by the atomic scattering factor $f_j(Q)$ which has an amplitude equal to the atomic number of j when Q equals zero. The behaviour of the atomic scattering factor is complicated when the wavelength of the incident X-rays is close to an absorption edge of the atom; this complication can be used to distinguish a chiral molecule from its mirror image using X-ray diffraction data from a crystal.

The resultant X-ray wave scattered by a molecule depends on the direction of $\underline{Q}$ as well as on its magnitude and can be represented by a complex number which is usually denoted by $F(\underline{Q})$. When Q equals zero, $F(\underline{Q})$ equals the total number of electrons in the molecule. At sufficiently low Q values it is possible to approximate $F(\underline{Q})$ by some geometrical shape which closely resembles the shape of the molecule when viewed at low resolution.

CHAPTER VI

Ideal Gases and Solutions

VI.1. Introduction

In an ideal gas there are no interactions between the molecules. This view of an ideal gas has its origins in the kinetic theory of gases. For a real gas to approach ideal behaviour, its concentration of molecules must be so low that they are unlikely to approach sufficiently closely for any appreciable interaction. This view may be extended to X-ray scattering. Then a gas is ideal if its molecular concentration is so low that no appreciable interference occurs between X-rays scattered by different molecules ie. intermolecular interference is negligible. Interference effects in the X-ray diffraction pattern of an ideal gas can only arise, by definition, between X-rays scattered in the same direction by different parts of the same molecule ie. all interference is intramolecular.

By analogy, an ideal solution contains solute molecules which can be considered to behave in the same way as the molecules in an ideal gas. If a real solution is to approach ideal behaviour, its solute molecules cannot interact appreciably either with each other or with solvent molecules. For the present purposes, a solution can be considered to be ideal if the X-ray scatter from the solute molecules is not modified by intermolecular interference. It turns out that a dilute solution in which the solute molecules are very much larger than the solvent molecules is ideal, in this sense, at low Q values. Thus, by analogy with an ideal gas, interference can only arise between X-rays scattered, in the same direction, by different parts of the same solute molecule.

Because interference between X-rays scattered by different molecules can be neglected, the scattering properties of ideal gases and solutions are particularly simple; they are, therefore, the first systems whose special properties are considered in this book. The absence of intermolecular interference effects makes it possible to deduce information about molecular structure without the need to consider the complication of intermolecular interference. Unfortunately the information which can be deduced is limited by the random orientations of the molecules in gases and solutions.

VI.2. Scattered intensity distribution

If a gas consisted of only a single stationary molecule, the intensity of the X-rays which it scattered would be given by

$$I(\underline{Q}) = F(\underline{Q})\ F^*(\underline{Q}) \qquad \text{[eqn. II.5]}$$

$$= \sum_j f_j(Q)\ \exp\ (i\ \underline{r}_j\cdot\underline{Q}) \sum_k f_k(Q)\ \exp\ (-\ i\ \underline{r}_k\cdot\underline{Q}) \qquad \text{[eqn. V. 4]}$$

$$= \sum_j \sum_k f_j(Q)\ f_k(Q)\ \exp\ (i\ [\underline{r}_j - \underline{r}_k]\cdot\underline{Q}) \qquad \text{(VI.1)}$$

where both summations are taken over all the atoms in the molecule. Here $\underline{r}_j$ is the position of the j th atom in the molecule, with respect to some arbitrarily chosen origin, and f_j is its atomic scattering factor.

Now suppose that the gas contains N of these same molecules. Interference effects between X-rays scattered by different molecules are considered to be negligible and so the resulting diffraction pattern is obtained by adding the intensity contribution from each. (Had interference not been neglected it would have been necessary to add the amplitudes of the X-ray waves scattered by each molecule - with due regard to phase differences.) Adding all these intensity contributions, the total is given, according to equation VI.1, by

$$I(\underline{Q}) = N \sum_j \sum_k f_j(Q)\ f_k(Q)\ \exp\ (i\ \underline{r}_{jk}\cdot\underline{Q}) \qquad \text{(VI.2)}$$

where $\underline{r}_{jk}$ is the vector separation between the j th and k th atoms in a single molecule.

The molecules in a gas have random orientations; this randomness was ignored in the derivation of equation VI.2. Since the angles which vectors like $\underline{r}_{jk}$ make with $\underline{Q}$ depend on the orientation of the molecule, so does the value of $\underline{r}_{jk}\cdot\underline{Q}$ and, hence, $I(\underline{Q})$. At an instant in time, the contribution which each molecule will make to $I(\underline{Q})$ then depends on its orientation at that time. Since a gas contains very many molecules, every possible orientation will be represented at that instant. If the intensity contribution from each of these orientations is added, the resultant intensity will be characteristic of a spherically averaged molecule. Furthermore the contributions for each instant in time have to be summed over the time period during which the diffraction pattern is recorded. During this period each molecule is likely to assume every possible orientation - which also leads to the scattered X-rays conveying information about a molecule which has been spherically averaged.

Thus equation VI.2 has to be spherically averaged if it is to represent the intensity of the X-rays scattered by molecules in the gas phase. This equation is analogous to the Fourier transform as defined in equation II.6 - except that the summation of equation VI.2 is over a set of atoms rather than over infinitesimal scattering elements. By analogy with equation II.8, which is the Fourier transform of a function with spherical symmetry, equation VI.2 has then to be replaced by

$$I(Q) = N \sum_j \sum_k f_j(Q)\ f_k(Q)\ \text{sinc}\ (Qr_{jk}) \qquad \text{(VI.3)}$$

The factor $4\pi r^2$ appears only in the integral formulation - it represents the surface area of a sphere of radius r and generates a volume element, $d\underline{r}$, in equation

II.6, from an elemental length, dr, in equation II.8. Following the convention of this book, r_{jk} is the modulus of $\underline{r}_{jk}$ ie. it is simply the distance between the j th and k th atoms in a molecule.

According to equation VI.3, the intensity of X-rays scattered by a gas is independent of the direction of $\underline{Q}$; thus, according to Section VI.2, its diffraction pattern has circular symmetry - like the example of Fig. IV.3. Then I(Q) can be obtained by measuring the intensity distribution along a radius of the pattern and converting radial distance into a distance in Q-space by the methods of Section IV.3. Note that equation VI.3 contains "self terms" which arise when j equals k, and "cross terms" which arise when they are not equal. Since r_{jk} is equal to r_{kj}, the cross term between j and k is identical to that between k and j. These properties are often emphasised by writing equation VI.3 in the form

$$I(Q) = N \left\{ \sum_j f_j^2(Q) + 2 \sum_{j} \sum_{> k} f_j(Q)\, f_k(Q)\, \text{sinc}\,(Qr_{jk}) \right\}$$

Equation VI.3 can be used to determine the structures of molecules in ideal gases. It is particularly easy to understand how these structures can be determined by trial-and-error. A model has first to be proposed for the molecular structure; I(Q) is then calculated from the atomic scattering factors and coordinates of the model using equation VI.3. The result of this calculation is compared with the experimentally determined intensity distribution and, if necessary, the model is adjusted to improve the agreement between the observed and calculated results.

In practice electron diffraction is generally used for this purpose because electrons are more strongly scattered than X-rays. For most states of matter electrons are so strongly absorbed that very thin specimens have to be used - the need to produce very thin specimens can be a distinct disadvantage of electron diffraction. Absorption is much less of a problem for a gas because of the low concentration of absorbing molecules. Note that for electron diffraction a high vacuum is required throughout most of the apparatus to eliminate scatter by air - the need for a high vacuum is usually a further disadvantage of the technique.

This application can be used to introduce the idea of a scale factor which usually arises when interpreting X-ray diffraction patterns from real systems. It is not usually worthwhile to attempt to record I($\underline{Q}$) on an absolute scale by comparing it with I(O). Some indication of the difficulty of measuring I(O), in general, is given in Section III.7. If I(Q) of equation VI.3 is measured on an arbitrary scale, N cannot be determined and so it acts as an empirical scale factor. Then its value must be adjusted to obtain the best agreement between an intensity distribution calculated from a model and that obtained experimentally.

Note that, when Q is very close to zero, the theory used in this chapter breaks down. According to Section V.4, the scattered X-rays will convey information about very large distances within the specimen at these very low Q values. They will then be sensitive to distances between molecules ie. intermolecular interference will be detectable. The complete theory is given by Guinier and by James (see BIBLIOGRAPHY Section 2). However, for dilute gases and macromolecular solutions, the values of Q at which intermolecular interference is detectable are so close to zero that they are not usually measurable - therefore this effect will be neglected in the rest of the chapter.

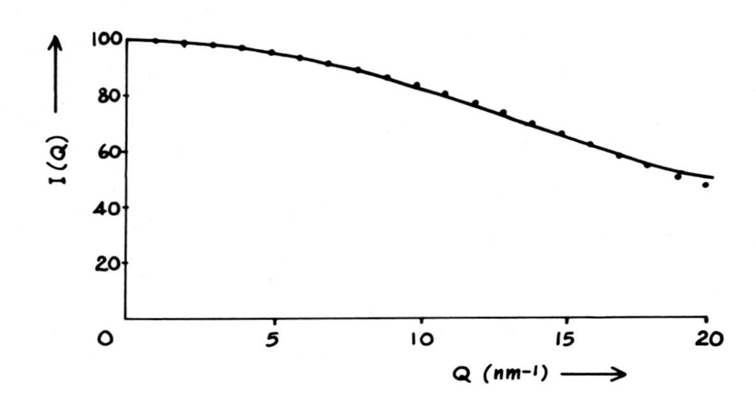

Fig. VI.1. Comparison of I(Q) for a spherically averaged water molecule calculated using equation VI.3 (continuous line) and equation VI.6 (dotted line)

VI.3. Intensity at low Q values

At low Q values the diffraction pattern is insensitive to details of molecular structure and the calculation of I(Q) can be simplified. This insensitivity was discussed in Section V.4. There the Fourier transform of a sphere, of radius a, was shown to be independent of the direction of $\underline{Q}$ and to be given by

$$F(Q) = n\ \Phi(Qa) \tag{VI.4}$$

where $\Phi(x)$ is defined in equation V.5 and the sphere contains n electrons. Now F(Q) already has spherical symmetry and is real; according to Section II.3 the intensity scattered by the sphere is $F^2(Q)$. The intensity scattered by an ideal gas of N spheres is then obtained by adding the intensity contribution from each ie.

$$I(Q) = NF^2(Q) \tag{VI.5}$$

Equation VI.5 could be considered as an alternative representation of equation VI.3 where F(Q) represents the Fourier transform of a spherically averaged molecule. Here we are considering the low resolution case where the molecules can be treated as spheres so that, from equations VI.4 and VI.5

$$I(Q) = Nn^2\ \Phi^2(Qa) \tag{VI.6}$$

Figure VI.1 compares the predictions of equations VI.3 and VI.6, at low Q values, for a water molecule in the ideal gas phase. The difference between these predictions is negligible at sufficiently low Q values. For calculating scatter from an assembly of molecules with random orientations, the Fourier transform of a sphere can be a particularly useful function. In the example shown in Fig. VI.1, the sphere was assigned a radius of 96 pm which is comparable with the length of an O-H bond.

VI.4. Solutions of macromolecules

The purpose of this section is to show that dilute solutions of macromolecules approach ideal behaviour at low Q values ie. the X-rays scattered by the solute

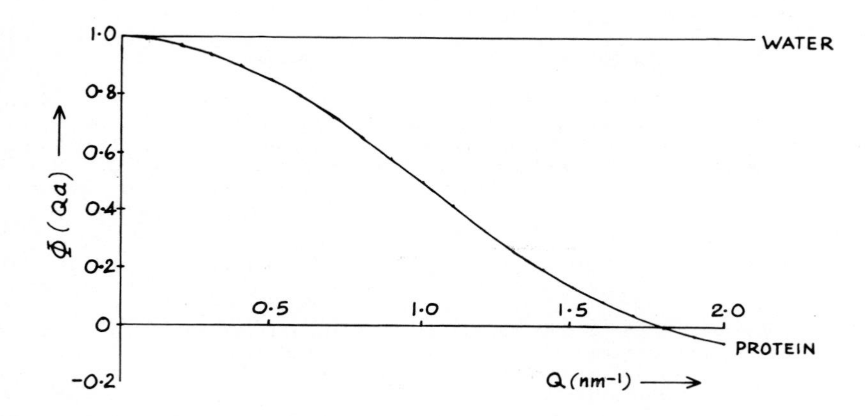

Fig. VI.2. Comparison of $\Phi(Qa)$ for spheres the size of protein and water molecules.

molecules do not interfere with those scattered by other solute or solvent molecules. Clearly interference with X-rays scattered by other solute molecules will be negligible if the solute is sufficiently dilute. But how can interference with X-rays scattered by the surrounding solvent molecules be ignored?

Suppose that the solute molecules in the solution are very much larger than the solvent molecules. For low values of Q, $F(Q)$, and hence $I(Q)$, will be insensitive to the detailed structure of the solution for the reasons given in Section V.4. If Q is restricted to sufficiently low Q values, $I(Q)$ will be sensitive to the dimensions of the large solute molecules without being sensitive to the relatively small distances between the solvent molecules. Thus the scattering from the solvent can be considered merely as a featureless background which has to be subtracted from the experimentally determined intensity distribution to obtain $I(Q)$ for the solute molecules alone.

How can we determine, experimentally, whether a solution is ideal? If $I(Q)$ is modified by intermolecular interference, its form will depend on the concentration of the solution. Thus, when $I(Q)$ is measured from a solution which is presumed to be ideal, the experiment should be repeated at a different concentration. The magnitude of the intensity distribution will change because the number of scatterers has changed. But if its shape is unchanged ie. the distributions are related simply by a scale factor - then the presumption of ideality is confirmed.

Figure VI.2 illustrates an example of the insensitivity of $F(Q)$ to detail, if it is restricted to low Q values. Here $\Phi(Qa)$ of equation VI.4 is plotted for spheres whose radii are 2.5 nm and 0.096 nm. The former could be considered as protein molecules and the latter as water molecules in which they are dissolved; Fig. VI.1 shows that a sphere of this latter radius is a good model for calculating the scatter from water molecules at these low Q values. N and n^2, in equation VI.6, can be ignored, for the present, as they simply adjust the relative scale of the two curves. Although the Fourier transform of the larger sphere falls off fairly rapidly, in the region of Q-space shown, that of the smaller sphere is effectively flat and featureless.

VI.5. Deduction of molecular structure

The structures of small molecules in ideal gases can often be solved using the Patterson function of Section III.3. According to Section VI.2, the intensity of X-rays scattered by an ideal gas or solution provides information about a spherically averaged molecule and is independent of the direction of $\underline{Q}$. The Patterson function is the Fourier transform of I(Q) and, according to equations II.8 and III.4, it can therefore be defined, for our present purposes by

$$P(r) = \int_0^\infty 4\pi Q^2 \, I(Q) \, \text{sinc}\,(Qr) \, dQ \qquad \text{(VI.7)}$$

Now P(r) is independent of direction and is a one-dimensional function. Thus peaks in P(r), computed from I(Q), now represent distances rather than vector separations. If I(Q) is measured to sufficiently high values of Q, P(r) will then yield interatomic distances. Each peak will represent the distance between two atoms in a molecule - distances between atoms in different molecules are very great, ie. they would appear at very high r values, which, in any case, would not be detectable because I(Q) is independent of intermolecular interference effects for an ideal gas. The height is related to the product of the electron densities of the two atoms whose interatomic distance the peak represents - as explained in Section III.3. For small molecules, containing only a few atoms, the structure can then be reconstructed from these distances. This method is confined to molecules in the gas phase because solutions are only ideal at low Q values and then I(Q) is insensitive to details of molecular structure.

Deduction of molecular structure has two potential disadvantages: both arise from Fourier transformation of experimental data - P(r) of equation VI.7 is a Fourier transform. According to this equation, P(r) is an integral which has to be evaluated between the limits of zero and infinity. In practice, I(Q) can only be measured over a finite range of Q-space, as described in Section III.6. Furthermore it was shown in this section that, if Q is too restricted in range, spurious peaks can arise in P(r). The second disadvantage arises because measurement of I(Q) is subject to experimental noise. Fourier transformation is very sensitive to noise which, in this case, could distort the appearance of P(r). For this reason, when a molecular structure has been solved by deduction, the expected intensity distribution should always be computed from its coordinates and compared with that observed - as a check on its validity. Although this calculation also involves Fourier transformation, the function which is transformed is now a set of model parameters with no experimental noise.

The structure of a spherical scatterer can be deduced by the direct method of Section III.4; this approach can sometimes be usefully applied to macromolecules in ideal solutions. Figure VI.3 shows that I(Q) for a sphere passes through a series of peaks separated by zero values of the intensity. The structure of a sphere is independent of the direction of a vector, $\underline{r}$, in real space and its Fourier transform is independent of the direction of $\underline{Q}$ as described for a spherically symmetric function in Section II.4; the problem of determining the structure of a spherical scatterer is thus identical to the example of Section III.4. Other spherically averaged shapes closely resemble spheres at low Q values - so does the method work for them too? Suppose I(Q) is carefully measured and the necessary corrections applied. If it passes smoothly through zero between peaks, the spherically averaged scatterer closely resembles a sphere at this resolution. The

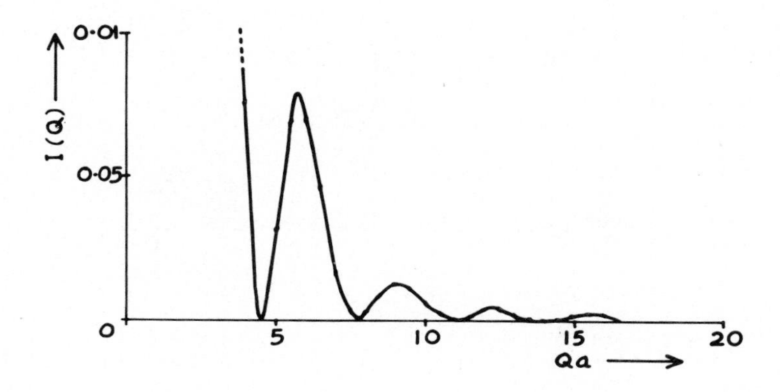

Fig. VI.3. I(Q) for a sphere of radius a (on an arbitrary scale)

method can then be applied. If I(Q) does not pass through zero neither does F(Q) - and the method breaks down. Note that this approach ignores intermolecular interference, ie. it is invalid for non-ideal solutions, and is subject to the potential disadvantages of the previous paragraph - resulting models should, therefore, be checked in the same way.

VI.6. Guinier's law

Guinier's law provides a method for deducing some information on the structure of solute molecules in ideal solutions - whatever their shape. Note that the law applies only to ideal solutions since its derivation ignores interference between X-rays scattered by different molecules. Thus its validity is restricted to dilute solutions of macromolecules and application to concentrated solutions, gels and fibres is invalid. Furthermore it is specifically derived for very low Q values at which diffraction data are insensitive to details of molecular structure. Readers who are not interested in the derivation of the law should skip the next five paragraphs.

In Fig. VI.4 a beam of X-rays is incident on a scattering body which is presumed to be in an ideal solution. The reason for this presumption is that, if diffraction effects at very low Q values are to provide information on its structure, the body must be at least as large as a macromolecule - it cannot, therefore, exist in the gas phase and can only attain the ideal behaviour of this chapter in a dilute solution. Consider the scattered beam shown in the figure; the ℓ-axis is defined to be perpendicular to the incident beam in the plane defined by the incident and scattered beams. Now $\underline{Q}$ terminates at the surface of the Ewald sphere (Section IV.2) which, at very low Q values, is effectively a plane which is perpen-

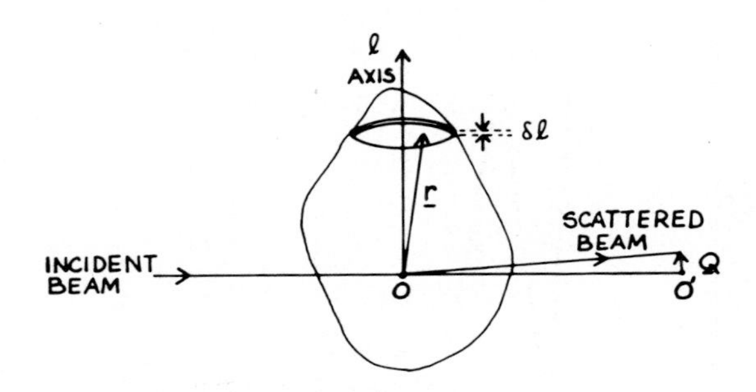

Fig. VI.4. Figure for the derivation of Guinier's law.

dicular to the incident beam because such a small fraction of its surface is being considered. Thus $\underline{Q}$ is roughly perpendicular to the incident beam and, therefore, parallel to the ℓ-axis.

At low Q values the Fourier transform of the body will be insensitive to the fluctuations in the electron density. Thus the electron density of the body is effectively constant and the Fourier transform becomes

$$F(\underline{Q}) = \rho \int \exp\,(i\;\underline{r}.\underline{Q})\;d\underline{r} \tag{VI.8}$$

where the integration is taken over the whole body and $\underline{r}$ represents a point within it, with respect to an origin O. Since the ℓ-axis is nearly parallel to $\underline{Q}$

$$\underline{r}.\underline{Q} = \ell Q \tag{VI.9}$$

where ℓ is the projection of $\underline{r}$ on to the ℓ-axis. If the result of equation VI.9 were to be substituted into equation VI.8, it would be necessary to integrate with respect to ℓ. Now, if $\delta\underline{r}$ is a volume element in the body, it can be expressed, in terms of ℓ, as

$$\delta\underline{r} = \sigma(\ell)\;\delta\ell \tag{VI.10}$$

where $\sigma(\ell)$ is the cross-sectional area of the body at a distance ℓ from O and $\delta\ell$ is an elemental distance as in Fig. VI.4. From equations VI.8, VI.9 and VI.10, the Fourier transform of the body becomes

$$F(Q) = \rho \int \sigma(\ell)\;\exp\,(i\;\ell\;Q)\;d\ell \tag{VI.11}$$

Note that equation VI.11 is not concerned with the direction of $\underline{Q}$ because this vector is now considered to be coplanar with the ℓ-axis as well as with the incident and scattered beams ie. the problem is one-dimensional.

Equation VI.11 can be simplified further. This simplification is achieved by noting that

$$\exp(x) = 1 + x + (x^2/2) + - - -$$

Since Q is very small we can neglect higher terms when expanding equation VI.11 and the result is

$$F(Q) = \rho \int \sigma(\ell)\, d\ell + i\, Q\, \rho \int \sigma(\ell)\, \ell\, d\ell - (Q^2\rho/2) \int \sigma(\ell)\, \ell^2\, d\ell \qquad \text{(VI.12)}$$

If the origin, O, of the body is defined to be at its centroid

$$\int \sigma(\ell)\, \ell\, d\ell = 0 \qquad \text{(VI.13)}$$

since the centroid is defined as the point about which the integral over all the elemental first moments is zero. The volume, V, of the body is given by

$$V = \int \sigma(\ell)\, d\ell \qquad \text{(VI.14)}$$

and a function, G_ℓ, is defined by

$$G_\ell^2 = (1/V) \int \ell^2\, \sigma(\ell)\, d\ell \qquad \text{(VI.15)}$$

From equations VI.12 to VI.15

$$F(Q) = \rho V\, [1 - (Q^2 G_\ell^2/2)] \qquad \text{(VI.16)}$$

Now the [bracketed] function could be considered as the first two terms in the expansion of an exponential. If Q is so small that only the first three terms are required for the expansion of equation VI.11 into equation VI.12, Q^2 will be so much smaller that equation VI.16 can be rewritten in the form

$$F(Q) = \rho V \exp(- Q^2 G_\ell^2/2) \qquad \text{(VI.17)}$$

For an ideal solution, consisting of N such scattering bodies, the resultant intensity will be given by

$$I(Q) = NF^2(Q) \qquad \text{[eqn. VI.5]}$$

$$= N\rho^2 V^2 \exp(- Q^2 G_\ell^2) \qquad \text{[eqn. VI.17]}$$

$$= I(0) \exp(- Q^2 G_\ell^2) \qquad \text{(VI.18)}$$

Here I(O) is the value predicted for I(Q) when Q equals zero. It is introduced to avoid writing $N\rho^2V^2$ as well as to emphasise that, in practice, this composite term will act merely as an empirical scale factor, since I(Q) will not usually be measured on an absolute scale.

In a solution the scattering bodies will have all possible orientations and so equation VI.18 has to be spherically averaged. Since the only term in equation VI.18 which depends on the structure of the body is G_ℓ^2, this averaging can be achieved simply by replacing it with its spherical average - denoted by $< G_\ell^2 >$. It will now be shown that $< G_\ell^2 >$ is related to the radius of gyration of the body, which is defined by

$$G^2 = (1/V) \int r^2 \, d\underline{r} \qquad \text{(VI.19)}$$

Two further axes, p and q, have to be defined so that ℓ, p and q are mutually perpendicular. If ℓ, p and q are the projections of $\underline{r}$ on to these axes

$$G^2 = (1/V) \int (\ell^2 + p^2 + q^2) \, d\underline{r} \qquad \text{[eqn. VI.19]}$$

$$= G_\ell^2 + G_p^2 + G_q^2 \qquad \text{[eqns. VI.10 and VI.15]}$$

and therefore

$$G^2 = < G_\ell^2 > + < G_p^2 > + < G_q^2 > \qquad \text{(VI.20)}$$

since G^2, as defined in equation VI.19, is independent of the orientation of the body and is, consequently, identical to its spherical average. When the body has been spherically averaged its properties must be isotropic and so

$$< G_\ell^2 > = < G_p^2 > = < G_q^2 > \qquad \text{(VI.21)}$$

From equations VI.20 and VI.21

$$< G_\ell^2 > = G^2/3 \qquad \text{(VI.22)}$$

Substituting this result into equation VI.18 yields

$$I(Q) = I(0) \exp(- Q^2G^2/3) \qquad \text{(VI.23)}$$

Equation VI.23 is a statement of Guinier's law. Table VI.1 lists a few formulae which allow the radius of gyration, G, to be calculated for a few different shapes. How good an approximation is Guinier's law? Figure VI.5 compares an exact calculation of I(Q) for a sphere of radius 2.5 nm with the prediction of equation VI.23; readers who have followed the derivation should note that, for the purpose of this comparison, the scale factor

$$I(0) = Nn^2$$

was set equal to unity. Guinier's law provides a good approximation at low Q values.

In practice Guinier's law is generally used to deduce the radius of gyration of the molecules in a solution. Taking logarithms of both sides of equation VI.23 gives

$$\ell n \, [I(Q)] = - (Q^2G^2/3) + \ell n \, [I(0)]$$

Conventionally ℓn [I(Q)] is plotted against Q^2 - a "Guinier plot". The slope of the graph then yields a value for G - assuming that all the large scattering

TABLE VI.1. Radii of gyration for bodies of simple shape

Sphere	Radius a	$(3/5)^{\frac{1}{2}}$ a
Ellipsoid	Semi-axes a, a and γa	$\{(2+\gamma^2)/5\}^{\frac{1}{2}}$ a
Long, thin rod	Length ℓ	$\ell/12^{\frac{1}{2}}$
Random coil *	Mean-square end-to-end distance * $\overline{h^2}$	$(\overline{h^2})^{\frac{1}{2}}/6^{\frac{1}{2}}$

* See C. Tanford, "Physical Chemistry of Macromolecules," Wiley, 1961, pp. 150-154 where a "completely unrestricted polymer chain" is a "random coil".

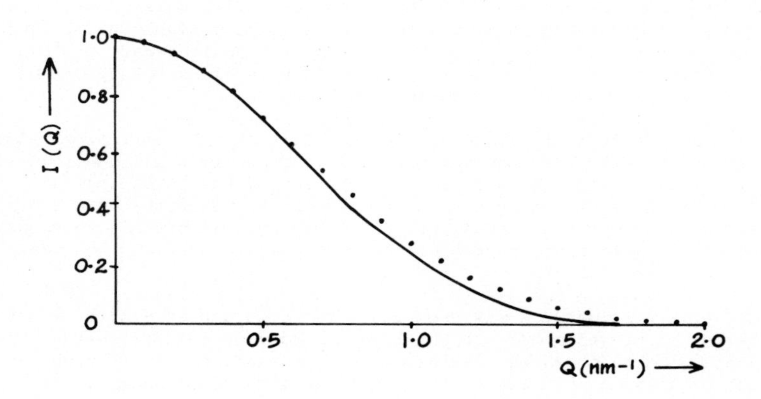

Fig. VI.5. Comparison of I(Q) for a sphere calculated using equation VI.6 (continuous line) and Guinier's law (dotted line).

bodies in the solution are identical. Note that all intensity measurements have to be corrected by subtracting the solvent contribution, for the reasons given in Section VI.4; this contribution can be measured in a separate experiment. Ignorance of the value of I(O) does not affect the determination of G.

Although Guinier's law allows the radius of gyration of a body to be deduced, its dimensions cannot be deduced unless its shape is known. The mathematical reason is that the relationship between G and the dimensions of a body depend on its shape; Guinier's law provides a value for G but no shape information. Physically this ignorance of shape arises because the experimental data provide information on a spherically averaged scatterer. At the best of times it is difficult to distinguish, eg., a sphere from a spherically averaged rod. For the low Q values, ie. low resolution, at which Guinier's law applies the distinction is not possible in the absence of further information.

VI.7. Smearing

It has already been indicated, in Section VI.2, that, in a gas, there are so few molecules that the intensity of scattered X-rays is weak; a similar problem arises in experiments on dilute solutions where there will be few solute molecules. In the case of solutions the problem cannot be solved by using electrons, as in Section VI.2, because the large number of solvent molecules would lead to considerable absorption. The only way to obtain a diffraction pattern in a reasonable time is to use as intense an X-ray beam as possible. Synchrotrons are increasingly being used as intense X-ray sources for investigating solutions but the more traditional approach is to use an X-ray beam with a large cross-sectional area.

How does increasing the cross-sectional area of the beam increase the scattered intensity? The intensity of X-rays from a conventional generator is dictated by the number of electrons striking the anode and is limited by the rate at which the anode can be cooled. If the electrons strike a large area the cooling will be more efficient and it is possible to generate a higher total intensity. But the cross-sectional area of the X-ray beam is also increased. Thus, although the X-ray intensity per unit area is limited, the total intensity can be increased by using a beam with a larger cross-sectional area.

The theory in this book has been derived for a highly collimated incident X-ray beam which meets the detector at a point of negligible dimensions; if a beam of larger cross-sectional area is used, a correction has to be applied. In practice a very narrow rectangular beam cross-section is generally used and a variety of correction procedures has been applied. The purpose of this section is not to catalogue these various procedures but to indicate, in general, how beam shape affects diffraction patterns.

A pin-hole collimated X-ray beam meets the detector at a point - but a beam with a larger cross-sectional area will be recorded as some intensity distribution, $h(\underline{x})$. This wide beam can be considered as a parallel bundle of highly collimated beams each of which gives rise to the same intensity distribution, $I(\underline{x})$, at the detector, but with respect to a slightly different origin; the origin is defined by the point where each collimated beam meets the detector as in Section IV.3. A wide beam, therefore, smears out the intensity distribution, $I(\underline{x})$, which would be expected from a collimated beam. According to Section II.4, the detected intensity, $D(\underline{x})$, is then given by

$$D(\underline{x}) = I(\underline{x}) \otimes h(\underline{x}) \qquad \text{(VI.24)}$$

In equation VI.24, $D(\underline{x})$ is measured and $h(\underline{x})$ is measurable: $I(\underline{x})$ is required to calculate $I(\underline{Q})$, by the methods of Section IV.3, in order to relate experiment to theory.

$I(\underline{x})$ can be recovered from $D(\underline{x})$ and $h(\underline{x})$ by reversing the process of convolution in equation VI.24 ie. by deconvolution. The need for deconvolution occurs in many areas of science and engineering - the method described here exploits the properties of the Fourier transform. If T denotes the operation of Fourier transformation, transforming both sides of equation VI.24 yields

$$T\{D(\underline{x})\} = T\{I(\underline{x}) \otimes h(\underline{x})\}$$

$$= T\{I(\underline{x})\}\, T\{h(\underline{x})\}$$

according to Section II.4. Rearranging and applying the inverse transform, T^{-1}, gives

$$T^{-1}\, T\{I(\underline{x})\} = T^{-1}\,[T\{D(\underline{x})\}/T\{h(\underline{x})\}]$$

which is simply

$$I(\underline{x}) = T^{-1}\,[T\{D(\underline{x})\}/T\{h(\underline{x})\}] \qquad \text{(VI.25)}$$

Equation VI.25 provides a method for recovering $I(\underline{x})$ from $D(\underline{x})$ if the intensity distribution of $h(\underline{x})$ is measured; it is not necessary to measure the intensity of either $h(\underline{x})$ or $D(\underline{x})$ on the same scale, as an absolute measurement of $I(\underline{x})$ will not usually be required.

Accurate representation of $h(\underline{x})$ may sometimes be complicated by the oblique incidence of the scattered X-rays. Suppose $h(\underline{x})$ is measured by detecting the undeviated X-ray beam on a flat film which is perpendicular to it - as in Fig. IV.4. Then $h(\underline{x})$ is the intensity distribution around the origin. If the beam were scattered without change of cross-sectional shape to some other point on the film it would not be incident at right angles. Then the area of film blackened by this scattered beam would not be exactly the same as that blackened by the undeviated beam; the shape obtained with the undeviated beam would be distorted, in the scattered case, by oblique incidence. This effect has to be allowed for if equation VI.25 is to be used to recover an accurate representation of $I(\underline{x})$.

VI.8. Summary

In an ideal gas interference between X-rays scattered by different molecules is considered to be negligible. Thus all interference effects arise between X-rays scattered by different parts of the same molecule. Since the molecules in a gas have random orientations, the diffraction pattern has circular symmetry and conveys information on the structure of a spherically averaged molecule.

A solution is considered to be ideal if the scattering properties of its solute molecules are identical to those of a molecule in an ideal gas. Dilute solutions of macromolecules are ideal, in this sense, at low Q values. Guinier's law provides a method for deducing the radius of gyration of solute molecules in an ideal solution. An X-ray beam of large cross-sectional area may be used, instead of a pin-hole collimated beam, to increase the intensity of the diffraction pattern - a correction then has to be applied to the intensity distribution at the detector before $I(Q)$ can be calculated.

CHAPTER VII

Liquids and Amorphous Solids

VII.1. Introduction

When X-rays are scattered by the molecules in a liquid or a solid, interference occurs between the waves scattered, in the same direction, by different molecules. Intermolecular interference is now important because the molecules are much closer together than they would be in an ideal gas. Consequently the theory for ideal gases has to be extended to include inter- as well as intra-molecular interference if it is to be applied to liquids and solids. In Chapters VIII and X it will be shown that this extension is particularly easy for crystalline solids because their molecules are regularly arranged in space. The molecules in liquids are irregularly arranged and have no preferred orientation. Amorphous solids, often called "glasses", are characterised by a similarly irregular arrangement of molecules and lack of molecular orientation. Nevertheless one approach to calculating intermolecular interference effects, for these irregular systems, treats them as disordered crystals - this is the "paracrystal theory for disorder of the second kind" as described, for example, by Hosemann (see BIBLIOGRAPHY Section 9). In complete contrast, the more modern approach, which is adopted here, actually exploits the irregularity which is characteristic of these systems for calculating intermolecular interference effects.

Liquids are much more like dense gases than disordered crystals. A crystal which is subjected to shear stress stores the deformation energy and returns to its original shape when the stress is removed - it is elastic. In contrast liquids and gases both flow when a shear stress is applied and do not store their deformation energy - they are viscous. Thus the interactions between molecules must be similar in liquids and gases, although in a crystalline solid they are very different. This distinction is emphasised by the symmetry changes which accompany phase transitions. When a crystal melts a loss of symmetry occurs but no symmetry change accompanies vapourisation of a liquid. Indeed the only clear distinction between a liquid and a dense gas is that the former is bounded by a surface that has one face which is not defined by a container.

Why consider liquids and amorphous solids together? Both are characterised by a seemingly haphazard arrangement of molecules - the only difference is that the molecules in a liquid are free to move around so that their positions are constantly changing. An X-ray diffraction pattern conveys information about the structure of a scatterer averaged over the time taken to record it. Therefore the technique cannot detect the changing positions of the molecules in a liquid but only their averaged, irregular positions. Consequently the same effects are observed in X-ray diffraction patterns from both liquids and amorphous solids.

X-Ray diffraction studies of these systems are usually performed to investigate how the constituent molecules are arranged in them. Since this arrangement is so irregular, it would be pointless to attempt to specify the position of each molecule - especially as there will be around 10^{23} molecules in a real sample! Thus the description of the molecular arrangement must be a statistical one.

VII.2. Intermolecular interference

Liquids and solids consist of an assembly of molecules in space. According to Section II.4 the Fourier transform of this assembly is then given by the sum of the transforms of the individual molecules. Remember that phase differences, introduced by each molecule having a different position, have to be taken into account - just as they did when atomic scattering factors were added to obtain a molecular transform in Section V.3. Taking into account these phase differences allows for interference between X-rays scattered by different molecules. In liquids and amorphous solids the molecules have random orientations. Following essentially the same arguments as in Section VI.2, the calculated intensity distribution is then given by

$$I(Q) = \sum_j \sum_k F_j(Q) F_k(Q) \text{ sinc } (Qr_{jk}) \qquad \text{(VII.1)}$$

Here $F_j(Q)$ is the spherically averaged Fourier transform of the j th molecule which is given by equation VI.3; at low Q values the approximate form of equation VI.6 may be applicable. The spherically averaged transform is appropriate here because, as in a gas, the molecules have random orientations - for further details see Section VI.2. (Note that, when calculating the spherically averaged molecular transform in Section VI.2, the atomic scattering factors did not have to be spherically averaged because, according to Section V.2, they already had spherical symmetry.)

The intensity distribution, $I(Q)$, of equation VII.1 is independent of the direction of $\underline{Q}$ and, hence, diffraction patterns from liquids and amorphous solids have circular symmetry for the reasons discussed in Section IV.2. Figures VII.1, VII.2 and VII.3 provide examples of such diffraction patterns. When the intensity distribution along a radius of one of these patterns is converted into Q-space, it gives $I(Q)$ for the sample. $I(Q)$ does not depend on the direction of $\underline{Q}$ because of the irregularity of the structure. A molecule will have some arrangement of molecules around it which, in total, might be expected to give rise to some scattered intensity distribution, $I(\underline{Q})$. Some other molecule will have the same arrangement of neighbours but with a different overall orientation - because there are so many molecules in a macroscopic sample. For the entire sample the diffraction pattern then provides information about the structure of such a molecular cluster averaged over all possible orientations; thus $I(Q)$ is independent of the direction of $\underline{Q}$ for essentially the same reason as described in Section VI.2 for an ideal gas.

Now that the applicability of equation VII.1 has been established, it may be manipulated into a more useful form. The first stage is to rewrite it as

$$I(Q) = \sum_j F_j^2(Q) + \sum_{j \neq} \sum_k F_j(Q) F_k(Q) \text{ sinc } (Qr_{jk})$$

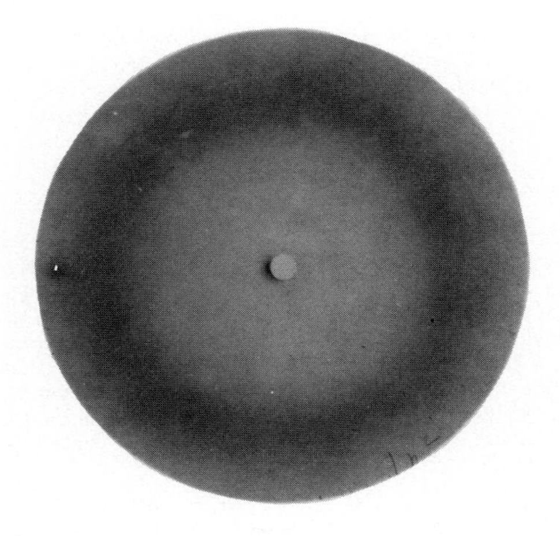

Fig. VII.1. X-Ray diffraction pattern of water (taken by D.S. Hickey).

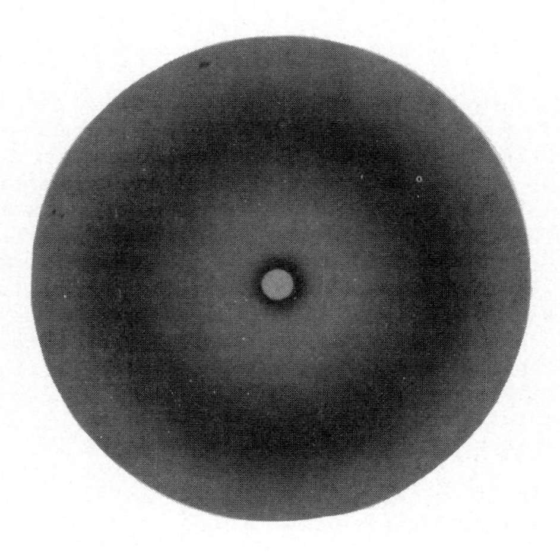

Fig. VII.2. X-Ray diffraction pattern of rubber (unstretched).

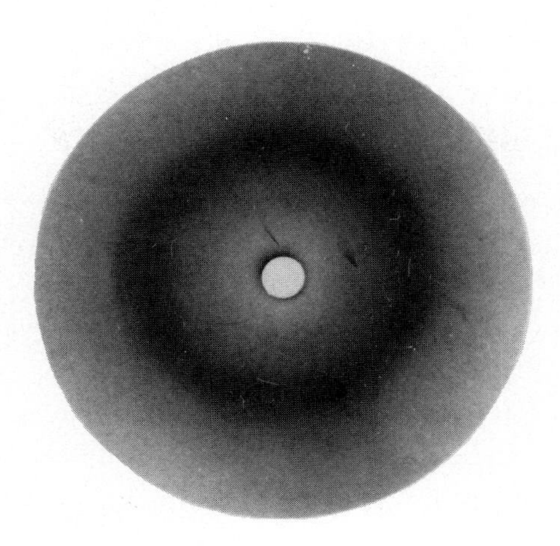

Fig. VII.3. X-Ray diffraction pattern of normal human costal cartilage.

If the gas consists of N identical molecules, this equation can be simplified to

$$\left.\begin{aligned} I(Q) &= NF^2(Q)S(Q) \\ S(Q) &= 1 + (1/N) \sum_{j} \sum_{\neq k} \operatorname{sinc}(Qr_{jk}) \end{aligned}\right\} \quad \text{(VII.2)}$$

Equation VII.2 is identical to equation VI.5 for an ideal gas - except for the appearance of S(Q). The only difference between the derivations of equations VI.5 and VII.2 is that the latter does not ignore intermolecular interference; therefore, S(Q) must allow for this effect and is called the interference function. (It is also sometimes called the "structure factor" but this nomenclature can be confusing because the term is used in a different way by crystallographers.)

Thus intermolecular interference effects can be represented by an interference function, S(Q); the next step is to derive a more useful form for S(Q) than that in equation VII.2. Suppose that every molecule in the sample could be considered to have an "average" relationship with its neighbours. Then each of the N molecules could interact with (N - 1) others and there would be N(N - 1) identical interactions. As a result, the definition of S(Q) in equation VII.2 could be rewritten as

$$\begin{aligned} S(Q) &= 1 + (N - 1) \sum_{k} \operatorname{sinc}(Qr_k) \\ &= 1 + N \sum_{k} \operatorname{sinc}(Qr_k) \end{aligned} \quad \text{(VII.3)}$$

The second line follows from the first since N and (N - 1) are essentially the same for an assembly of around 10^{23} molecules. Note that, in equation VII.3, r_k is the distance of the k th molecule from the centre of some arbitrarily chosen molecule. Thus an arbitrarily chosen molecular centre has become the origin of our coordinate system.

However every molecule in a liquid or amorphous solid does not have exactly the same relationship with its neighbours - thus the derivation of equation VII.3 was somewhat premature. The statistical nature of the structure can be allowed for properly by introducing the radial distribution function, $g(r)$, which is the probability of finding a molecule at some distance, r, from the origin. Consider an element, δr, of the radial distance, r; this elemental length defines an elemental volume of

$$4\pi r^2 \, \delta r$$

and the probability of finding each molecule within this volume is

$$\{4\pi r^2 \, \delta r/V\} \, g(r)$$

where V is the total volume of the liquid or amorphous solid. Now $g(r)$ does not exist only at a series of points - it is a continuous function. Thus the summation of equation VII.3 has to be replaced by an integral, now that a definite molecular position has been replaced by the probability of finding the molecule at that point. The result is

$$\left.\begin{aligned} S(Q) &= 1 + n_o \int 4\pi r^2 \, g(r) \, \text{sinc}\,(Qr) \, dr \\ n_o &= N/V \end{aligned}\right\} \quad \text{(VII.4)}$$

where n_o is the number of molecules per unit volume. The limits of integration are between zero and infinity if the liquid is considered to be of effectively infinite extent on the molecular scale.

In Section VII.4 it will be seen that $\{S(Q) - 1\}$ and $\{g(r) - 1\}$ are related by Fourier transformation. Anticipating this result, consider the effect of replacing $g(r)$ in equation VII.4 by

$$1 + \{g(r) - 1\}$$

which yields

$$S(Q) = 1 + n_o \int 4\pi r^2 \, \text{sinc}\,(Qr) \, dr + n_o \int 4\pi r^2 \, \{g(r) - 1\} \, \text{sinc}\,(Qr) \, dr \quad \text{(VII.5)}$$

The first integral in equation VII.5 is almost exactly the same as the integral which arose in the derivation of equation V.5 in Section V.5; here the constant n_o replaces the effectively constant electron density, ρ, which appeared previously. Also in equation VII.5 the upper limit of integration is infinity. Equation V.5 shows that the value of this integral then tends to zero - because as x in equation V.5 tends to infinity the value of $\Phi(x)$ tends to zero. Therefore the first integral in equation VII.5 can be neglected.

In summary, for a liquid or amorphous solid, the scattered intensity distribution is given by

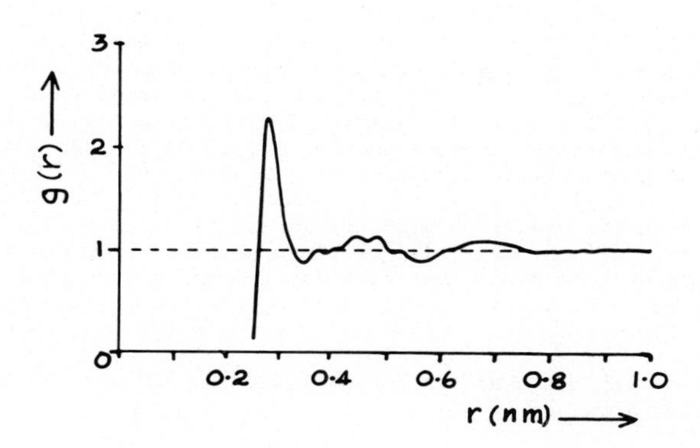

Fig. VII.4. Radial distribution function, g(r), of water. (The data used to produce this figure were taken from A.H. Narten and H.A. Levy, J.Chem. Phys. 55, 2263-2269, 1971.)

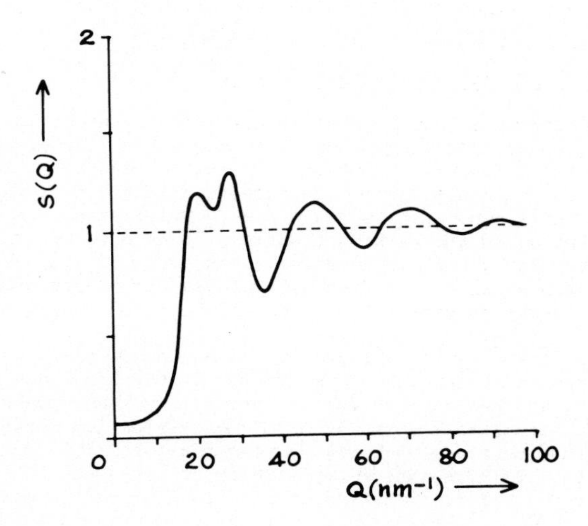

Fig. VII.5. Interference function, S(Q), for water. (The data used to produce this figure were taken from A.H. Narten and H.A. Levy, J.Chem.Phys. 55, 2263-2269, 1971.)

$$\left.\begin{aligned} I(Q) &= NF^2(Q)S(Q) \\ S(Q) &= 1 + n_o \int_o^\infty 4\pi r^2 \{g(r) - 1\} \operatorname{sinc}(Qr)\, dr \end{aligned}\right\} \quad \text{(VII.6)}$$

from equations VII.2 and VII.5. The interference function, $S(Q)$, allows for intermolecular interference effects and, not unexpectedly, depends on the radial distribution function, $g(r)$, which describes the statistical arrangement of molecules in the sample. Figure VII.4 shows an example of $g(r)$ for a liquid and Fig. VII.5 shows the corresponding interference function $S(Q)$. Both functions tend to values of unity - $g(r)$ at high r values and $S(Q)$ at high Q values.

Since $S(Q)$ contains peaks, so will the intensity distribution, $I(Q)$. Thus intermolecular interference gives rise to the appearance of peaks in a diffraction pattern. $S(Q)$ depends on the number of molecules per unit volume, n_o, but the dependence is rather complicated because $g(r)$ depends on n_o too - the overall effect and its influence on $I(Q)$ will be discussed in the next section.

The discussion in this next section will use the idea of the packing fraction, η, which is related to n_o by

$$\eta = (4/3)\pi a^3 n_o \quad \text{(VII.7)}$$

Here a is the radius of a spherical molecule ie. η is strictly defined only for spheres (in three dimensions). It is often convenient to think of the molecules in a liquid or solid as spheres and then η is the fraction of space actually occupied by molecules - as opposed to the space between them. Unlike n_o, η does not depend on the size of the molecules; consequently it is more suitable for a general discussion on the appearance of the diffraction pattern from a liquid or amorphous solid.

VII.3. Appearance of diffraction patterns

Interference effects appear in the intensity distribution, $I(Q)$, because the arrangement of molecules in a liquid, or amorphous solid, is not completely featureless. There will always be order of a statistical kind, ie. the arrangement of molecules is not truly random, because each molecule excludes the others from the space which it occupies itself. Figure VII.6 shows how the interference function, $S(Q)$, depends on the packing fraction, η, as a result of this "short-range order". The figure was computed for an assembly of spheres which only interact in so far as each sphere, of radius a, excludes the others from the volume which it occupies - the theory is given in Section VII.5.

The effect of interference is that $I(Q)$ is no longer given by the squared, spherically averaged molecular transform, $F^2(Q)$, of the ideal gas case, but is modulated by $S(Q)$ - it can then exhibit one or more fairly sharp peaks. Note, however, that the modulation leads to the retention of some intensity between these peaks. Figure VI.3 shows the relatively slowly varying form of $F^2(Q)$ for a sphere, on an arbitrary scale: Fig. VII.7 shows the appearance of $I(Q)$ when $F^2(Q)$ is multiplied by $S(Q)$ as in equation VII.6. Peaks appearing in $I(Q)$ can be more closely spaced than the peaks in $F^2(Q)$. The reason is that $F^2(Q)$ conveys information on molecular structure but $S(Q)$ conveys information on an arrangement of molecules. Since distances between molecules are much greater than the distances between atoms in a molecule, $S(Q)$ is a more oscillatory function than $F^2(Q)$ - according to Section V.4. Therefore peaks in $I(Q)$ can be more closely spaced than the peaks in $F^2(Q)$.

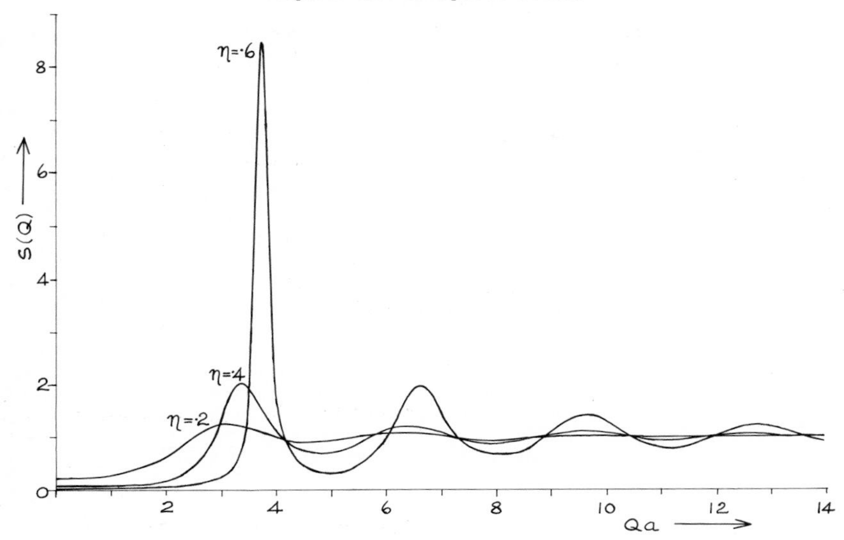

Fig. VII.6. Interference function, S(Q), for spheres of radius a at different values of the packing fraction, η.

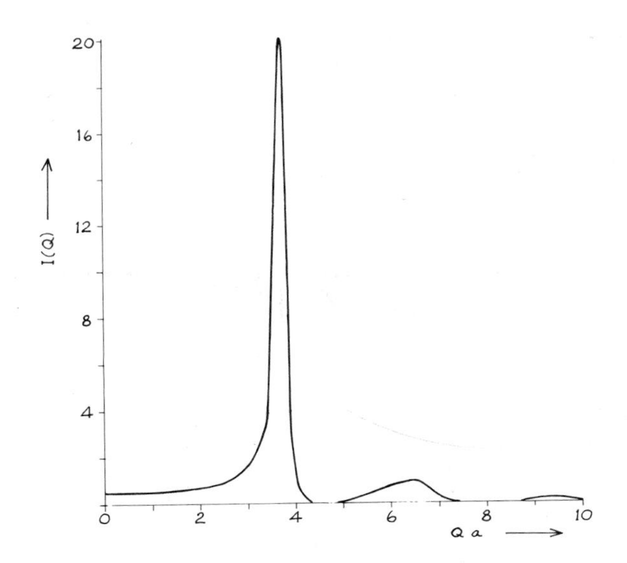

Fig. VII.7. Intensity distribution, I(Q), for spheres of radius a when the packing fraction, η, equals 0.6.

Figure VII.7 shows how successive peaks in I(Q) become progressively broader with increasing Q - this behaviour is characteristic of diffraction patterns from irregular assemblies of molecules. It arises from the properties of S(Q) whose peaks become progressively broader and less intense, as shown in Fig. VII.6. Consequently peaks in I(Q) tend to become less intense also, although this tendency can be modified by the behaviour of $F^2(Q)$.

Figure VII.6 shows that, as η increases, the peaks in S(Q), and hence in I(Q), become sharper; and the first peak, in particular, of S(Q) shifts to a slightly higher Q value. The sharpness arises because, as η increases, the molecules move closer together and interference effects become more marked. When η reaches its maximum value (0.74) the spheres will become close packed and the arrangement will be crystalline - we shall see later (in Section VIII.2) that S(Q), for a crystal, consists solely of a series of equally sharp peaks. At the other extreme, of very low η values, S(Q) has the form expected for an ideal gas, ie. by comparison of equations VI.5 and VII.6 it has a constant value of unity, and no peaks appear. Increased closeness also explains the peak shift when η increases. The closer together the molecules move, the higher the resolution of the diffraction data required to measure their separation ie. the information on their spacing is conveyed at higher Q values. Now that the diffraction pattern from a liquid or an amorphous solid can be recognised - how can g(r) be determined?

VII.4. Determination of the arrangement of molecules

It is always possible, at least in principle, to deduce the radial distribution function, g(r), which describes the arrangement of molecules in a liquid or amorphous solid, from the scattered intensity distribution, I(Q). The spherically averaged molecular transform, F(Q), can be calculated, presuming that the structure of the molecules is known, using equation VI.3. Division of I(Q), which is measured experimentally, by $F^2(Q)$ yields S(Q) according to equation VII.6. The effect of N can be neglected as it acts simply as a scale factor. Application of the inversion property of the Fourier transform, described in Section II.4, to the definition of S(Q) in equation VII.6 yields

$$g(r) - 1 = (1/n_o) \int 4\pi Q^2 \{S(Q) - 1\} \operatorname{sinc}(Qr)\, dQ \qquad \text{(VII.8)}$$

The limits of integration should be between zero and infinity but, according to Section III.6, there is a limit to the extent of Q-space which is measurable. In practice, therefore, the upper limit has to be truncated. No phase problem arises in the inversion of equation VII.6 because S(Q) is real and always positive - like I(Q) and $F^2(Q)$ from which it is calculated. Although S(Q) and, hence, the result from equation VII.8 will usually be obtained on an arbitrary scale, it is possible to obtain absolute values for g(r). It is only necessary to scale the result obtained for {g(r) - 1} so that g(r) tends to unity at high r values, to achieve the required result.

Deduction of g(r) in this way has two disadvantages: spurious peaks can appear in g(r), if an insufficient range of Q-space is used in equation VII.8, and the result is very sensitive to experimental noise. These disadvantages were discussed in Section VI.5. The advantage of the deductive approach is that it can yield a, possibly imperfect, trial model for g(r), if no other is available, as an aid to solution by trial-and-error.

Trial-and-error determination of g(r) does not involve Fourier transformation of noisy data and the truncation effects should not arise for the reason given in Section III.6. A trial model for g(r) is used to compute S(Q). Then the "experimental" form of $S(Q)$ is obtained from the measured intensity distribution, $I(Q)$, following division by $F^2(Q)$ - as before. Calculated and experimental forms of $S(Q)$ are compared as a test of the trial model. Alternatively the calculated form of $S(Q)$ can be multiplied by $F^2(Q)$ to yield a predicted $I(Q)$ which is compared with the experimental distribution to test the model. Further details on this approach to determining structures were given in Section III.5.

VII.5. Theories of g(r) for liquids

X-Ray diffraction data from liquids are often used to test theories of the liquid state; the purpose of this section is to indicate how these theories are related to $g(r)$ and thence, by equation VII.6, to $S(Q)$. It is not intended to give a thorough account of liquid state physics or even to justify all of the statements made; both would be far beyond the scope of this book. The interested reader is referred to the book by Croxton (see BIBLIOGRAPHY Section 9) for further details.

The total correlation function, $h(r)$ is defined by

$$h(r) = g(r) - 1 \qquad \text{(VII.9)}$$

Figure VII.4 shows that $h(r)$ then represents fluctuations of $g(r)$ about the value which it approaches at large r values. In order to proceed further, consider how one molecule influences the position of another. The first molecule can interact directly with the second but it can also influence the position of a third molecule - the position of the third then exerts a further influence on the position of the second. Thus the influence of one molecule on another can be either direct or indirect. This idea leads to the relationship

$$h(r) = C(r) + n_o\, h(r) \otimes C(r) \qquad \text{(VII.10)}$$

which is known as the Ornstein-Zernike equation and can be used as a definition of the direct correlation function, $C(r)$.

There is a simple relationship between the interference function, $S(Q)$, and the direct correlation function. Remember that the Fourier transform of the convolution of two functions is equal to the product of their individual Fourier transforms - as in Section II.4. Fourier transformation of equation VII.10 yields

$$\tilde{h}(Q) = \tilde{C}(Q) + n_o\, \tilde{h}(Q)\tilde{C}(Q) \qquad \text{(VII.11)}$$

where $\tilde{h}(Q)$ and $\tilde{C}(Q)$ are the transforms of $h(r)$ and $C(r)$, respectively. From equations VII.6, VII.9 and VII.11 it follows that

$$S(Q) = \{1 - n_o\, \tilde{C}(Q)\}^{-1} \qquad \text{(VII.12)}$$

Equation VII.12 allows $S(Q)$ to be calculated from the Fourier transform of $C(r)$ if n_o is known.

The direct correlation function, $C(r)$, can be related to the intermolecular potential function, $\phi(r)$, which describes the interaction between two molecules. At large values of r, strictly as r tends to infinity, $C(r)$ is given by

$$C(r) = -\,\phi(r)/k_BT \qquad \text{(VII.13)}$$

where k_B is Boltzmann's constant and T is the temperature, on the Kelvin scale, as

TABLE VII.1. Percus-Yevick Hard-Sphere Direct Correlation Function

$$C(r) = \begin{cases} \alpha + \beta(r/2a) + \gamma(r/2a)^3 & r < 2a \\ 0 & r \geqslant 2a \end{cases}$$

$$\alpha = -(1 + 2\eta)^2(1 - \eta)^{-4}$$

$$\beta = 6\eta\,\{1 + (\eta/2)\}^2\,(1 - \eta)^{-4}$$

$$\gamma = -(\eta/2)(1 + 2\eta)^2(1 - \eta)^{-4}$$

in Section II.6. A theory of the liquid state amounts to an expression for $\phi(r)$ - which at large r values is simply related to C(r) by equation VII.13. According to equation VII.12, S(Q) can be predicted from C(r) - such a prediction can be tested by comparison with experiment as described in Section VII.4.

A surprisingly successful agreement between theory and experiment can often be achieved by assuming that the only interaction between molecules is mutual exclusion - each excludes the others from the volume it occupies. Then $\phi(r)$ is given by

$$\phi(r) = \begin{cases} 0 & r \geqslant 2a \\ \infty & r < 2a \end{cases} \qquad \text{(VII.14)}$$

where the molecules are considered to be spheres of radius a. Equation VII.14 represents a "hard sphere" model for the van der Waals' interaction between two identical molecules. Better functions are available to represent van der Waals' interactions but they differ mainly in including a small, long-range attractive term. Long-range forces do not appear to contribute much to the structures of most liquids - although for water, long-range dipole-dipole interactions have to be taken into account if S(Q) is to be explained satisfactorily.

Combination of the hard sphere model with a very successful approximate theory (the Percus-Yevick approximation) allows S(Q) to be computed readily. The Percus-Yevick approximation neglects certain kinds of interactions between molecules. A polynomial representation for C(r) can then be developed which, according to equations VII.13 and VII.14 has the property that

$$C(r) = 0 \qquad r \geqslant 2a \qquad \text{(VII.15)}$$

The polynomial is given in Table VII.1; unlike the expression for C(r) in equation VII.13, this polynomial representation is valid for all r values. It follows from equations II.8 and VII.15 that the Fourier transform of C(r) is given by

$$\tilde{C}(Q) = \int_0^{2a} 4\pi r^2\; C(r)\; \text{sinc}\;(Qr)\; dr \qquad \text{(VII.16)}$$

since C(r) must be the same in all directions ie. it has spherical symmetry. If η and a are known, C(r) can be obtained using the equations in Table VII.1. Then

$\tilde{C}(Q)$ and n_o can be calculated from equations VII.16 and VII.7 respectively. Finally $S(Q)$ can be calculated from $\tilde{C}(Q)$ and n_o using equation VII.12.

VII.6. Summary

Liquids and amorphous solids have an irregular arrangement of molecules. But, in contrast to ideal gases, these molecules are sufficiently close for interference to occur between X-rays scattered in the same direction by different molecules. Thus the intensity distribution, I(Q), expected of an ideal gas, is modulated by an interference function, S(Q). Because the molecules have random orientations, I(Q) is independent of the direction of $\underline{Q}$ and the diffraction pattern has circular symmetry. Diffraction patterns from liquids and amorphous solids can be recognised because peaks in I(Q) become increasingly broader with increasing Q.

The interference function, S(Q), can be calculated from the structure of these irregular systems. Such structures are described by a radial distribution function, g(r), which is the probability of finding another molecule at a given distance from an arbitrarily chosen molecular centre. S(Q) is related to g(r) by a Fourier transformation. If the molecular structure is known, S(Q) can be calculated from I(Q) - g(r) can then be obtained by inverse transformation. Because of the errors which may be introduced by inverse transformation, it is preferable to develop a model for g(r), calculate S(Q) and compare the result with that observed. For liquids, theories have been developed which relate g(r), and hence S(Q), to intermolecular potential functions.

CHAPTER VIII

One-dimensional Crystals

VIII.1. Introduction

A crystal is a solid with a regularly repetitive structure. As in any solid, interference will occur between X-rays scattered in the same direction by different molecules because these molecules are packed together so closely. The regularity of this packing makes the interference effects particularly easy to calculate.

A one-dimensional crystal has a regularly repeating structure in one direction but this regular repetition is not a feature of the structure in other directions. One-dimensional crystals are considered in this book for three different reasons. Firstly, the one-dimensional crystal provides a simple system for illustrating the properties of crystals in general eg. in Section VIII.6. Secondly, the theory of diffraction by a one-dimensional crystal provides an essential background to understanding diffraction by three-dimensional crystals in Chapter X. Thirdly, one-dimensional crystals actually occur and their diffraction patterns have to be interpreted in practice. Some examples of one-dimensional crystals will be encountered in Chapter IX.

VIII.2. Interference effects

Figure VIII.1 shows a schematic example of a one-dimensional crystal; it consists of layers of bodies whose arrangement repeats regularly in one direction but not in others. In reality such a structure would be three-dimensional with the lines of bodies in the figure extending over planes which were perpendicular to the plane of the page - it is left to the reader's imagination to supply the third dimension. The bodies might be atoms or molecules or even structures made up of many molecules. The repeat direction will be designated the c-axis where c is the repeat distance. Figure VIII.2 shows how Fig. VIII.1 can be generated by the convolution of a layer of bodies with a row of points - as in Fig. II.5. This row of points is the one-dimensional lattice of the crystal.

According to Section II.4, the Fourier transform of a one-dimensional crystal is then given by the transform of a layer, $F(\underline{Q})$, multiplied by the transform of a one-dimensional lattice, $\psi(\underline{Q})$. Suppose that N points are required to generate the crystal. In order to calculate the Fourier transform of this row of discrete points, the integral of equation II.3 can be replaced by the summation

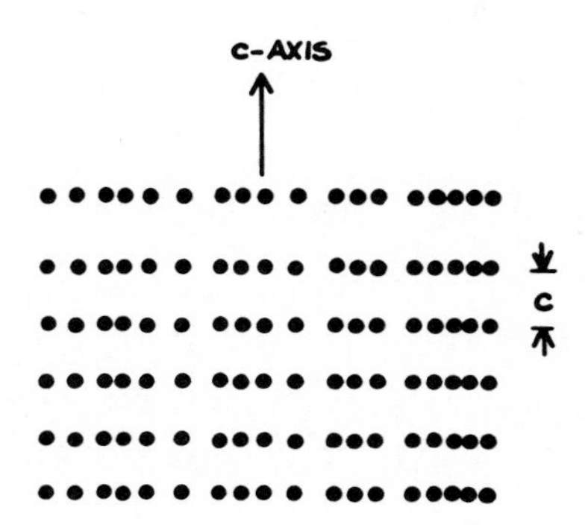

Fig. VIII.1. One-dimensional crystal.

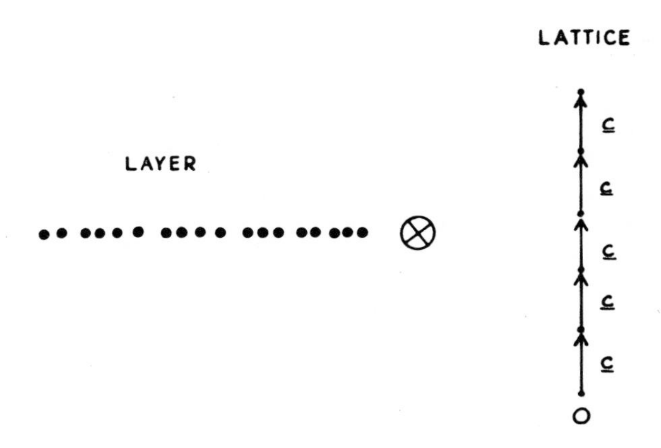

Fig. VIII.2. Generation of a one-dimensional crystal by convolution.

$$\psi(\underline{Q}) = \sum_{j=1}^{N} \exp\,(i\ \underline{r}_j . \underline{Q}) \tag{VIII.1}$$

where $\underline{r}_j$ is the position of the j th point with respect to some origin. In Fig. VIII.2 the origin is arbitrarily placed at the end of the lattice and each point is separated from its predecessor by a vector $\underline{c}$. Now, when equation VIII.1 is written out in full, it can be seen to be a geometrical progression which can be summed, in the usual way, as follows:

$$\psi(\underline{Q}) = 1 + \exp\,(i\ \underline{c}.\underline{Q}) + \exp\,(i\ 2\underline{c}.\underline{Q}) + - - -$$

$$- - - - + \exp\,(i\{N - 1\}\ \underline{c}.\underline{Q})$$

$$= \{1 - \exp\,(i\ N\underline{c}.\underline{Q})\}/\{1 - \exp\,(i\ \underline{c}.\underline{Q})\}$$

$$= \frac{\exp\,(i\ N\underline{c}.\underline{Q}/2)}{\exp\,(i\ \underline{c}.\underline{Q}/2)} \left\{ \frac{\exp\,(-\ i\ N\underline{c}.\underline{Q}/2) - \exp\,(i\ N\underline{c}.\underline{Q}/2)}{\exp\,(-\ i\ \underline{c}.\underline{Q}/2) - \exp\,(i\ \underline{c}.\underline{Q}/2)} \right\}$$

$$= \exp\,\{i(N - 1)\ \underline{c}.\underline{Q}/2\}\ \frac{\sin\,(N\underline{c}.\underline{Q}/2)}{\sin\,(\underline{c}.\underline{Q}/2)} \tag{VIII.2}$$

The intensity of X-rays scattered by the crystal can now be calculated. Since its Fourier transform is given by $F(\underline{Q})\psi(\underline{Q})$ the scattered intensity is given, according to equation II.5, by

$$I(\underline{Q}) = F(\underline{Q})\psi(\underline{Q})F^*(\underline{Q})\psi^*(\underline{Q}) \tag{VIII.3}$$

where $F^*(\underline{Q})$ is the complex conjugate of $F(\underline{Q})$. From equations VIII.2 and VIII.3

$$\left.\begin{aligned} I(\underline{Q}) &= S(\underline{Q})F(\underline{Q})F^*(\underline{Q}) \\ S(\underline{Q}) &= \sin^2\,(N\underline{c}.\underline{Q}/2)/\sin^2\,(\underline{c}.\underline{Q}/2) \end{aligned}\right\} \tag{VIII.4}$$

Here $S(\underline{Q})$ is the interference function for a one-dimensional crystal. Since the intensity scattered by a single layer in isolation would be $F(\underline{Q})F^*(\underline{Q})$ the interference function shows the effect of stacking N layers on top of each other, a distance c apart, to form a one-dimensionally periodic structure.

It can be shown analytically that, as N tends to infinity, $S(\underline{Q})$ tends to zero - except in those regions of Q-space which satisfy the equation

$$\underline{c}.\underline{Q} = 2\pi\ell \tag{VIII.5}$$

Here ℓ is any integer which can have positive, negative and zero values. Rather than deriving equation VIII.5 analytically, a numerical approach is adopted here. The analytical approach is described by Lipson and Taylor (see BIBLIOGRAPHY Section 10). In Fig. VIII.3, the interference function is plotted against Qc (ie. $\underline{c}.\underline{Q}$ when the two vectors are parallel) for different values of N. When N is small, the highest peaks in $S(\underline{Q})$ are regularly spaced a distance $2\pi/c$ apart in Q-space; these peaks all have the same height and width. As N increases the main peaks become very much higher and sharper. When N is large, $S(\underline{Q})$ consists of sharp peaks spaced $2\pi/c$ apart.

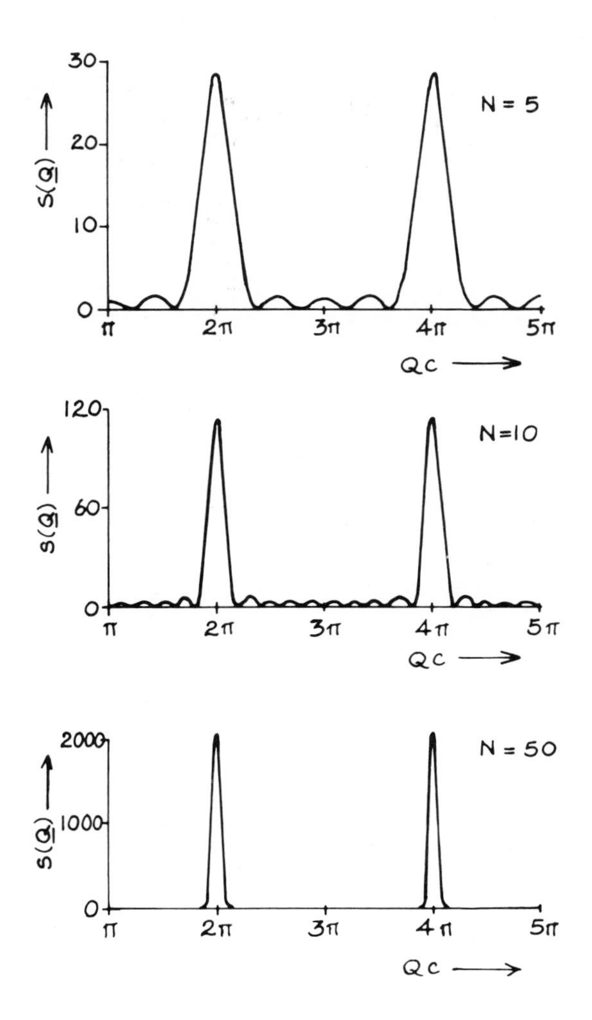

Fig. VIII.3. Crystal interference function, $S(\underline{Q})$, defined by equation VIII.4 as a function of Qc.

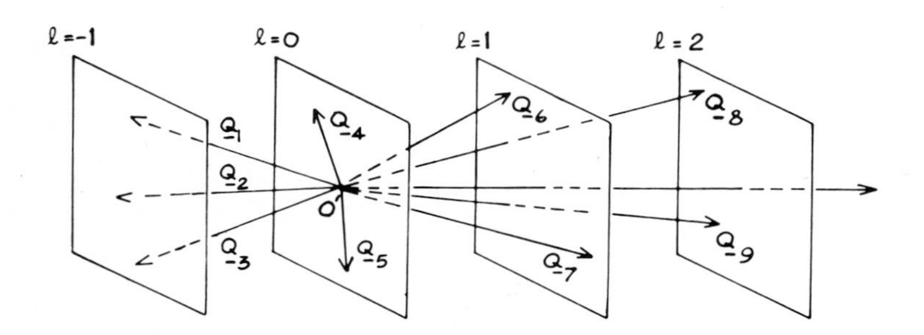

Fig. VIII.4. Equation VIII.5 defines a set of planes in Q-space which are perpendicular to the c-axis direction.

The one-dimensional regularity acts as a very unusual amplifier for the intensity which would be scattered by a single layer in isolation. From Fig. VIII.3 it appears that $S(\underline{Q})$ increases this intensity by an enormous factor, of the order of N^2, in a few selected regions of Q-space. But in the majority of Q-space all intensity is effectively obliterated. Because there is almost no scatter in most directions, this highly selective amplification does not violate the principle of conservation of energy - despite the magnitude of the gain.

This amplification has an important consequence, which is often overlooked, for systems which are mixtures of ordered and disordered phases. The amplified region of Q-space, for the ordered phase, will be very intense compared with the scatter from the disordered phase. In the relatively short time period required to measure the former, the latter may remain undetected. Thus X-ray diffraction is very insensitive to the presence of disordered phases in such systems.

The interference function for a crystal is very different from that for an irregular system such as a liquid or an amorphous solid. Figure VIII.3 shows that, for a crystal, the peaks are all equally sharp - in complete contrast to Fig. VII.6 where successive peaks increase in width. How does the sharpness of $S(\underline{Q})$ affect the appearance of the diffraction pattern? Before this question can be answered we shall have to consider how the diffraction pattern from a one-dimensional crystal is formed.

VIII.3. Formation of the diffraction pattern

It follows from equation VIII.5 that the intensity of X-rays, $I(\underline{Q})$, scattered by a one-dimensional crystal is non-zero only on a set of planes, spaced $2\pi/c$ apart in Q-space, which are perpendicular to the c-axis direction. The interference function, $S(\underline{Q})$, of equation VIII.4 is effectively non-zero only when equation VIII.5 is satisfied. Figure VIII.4 shows that equation VIII.5 defines a set of parallel planes perpendicular to the c-axis; remember that

$$\underline{c}\cdot\underline{Q} = \underline{Q}\cdot\underline{c}$$

and that the dot product of two vectors is the projection of the first on to the second - multiplied by the length of the second. Whenever ℓ is zero, $\underline{Q}$ must be

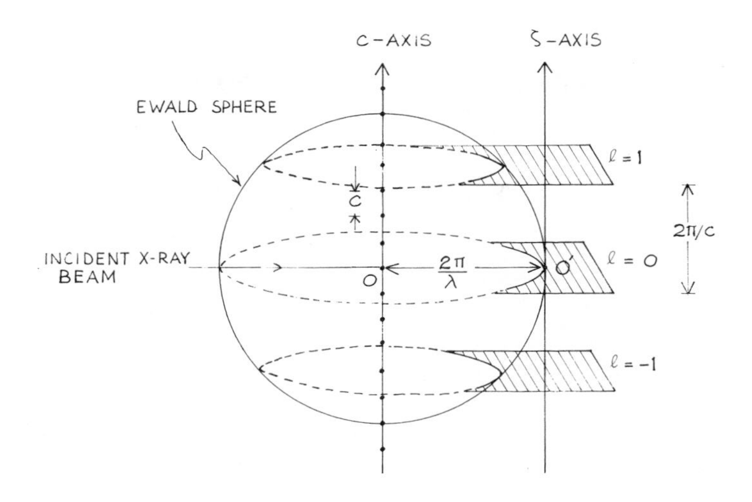

Fig. VIII.5. Formation of the diffraction pattern from a one-dimensional crystal.

perpendicular to $\underline{c}$ since only then can their dot product be zero; $\underline{Q}_4$ and $\underline{Q}_5$ in the figure provide examples. All such vectors lie on a plane perpendicular to $\underline{c}$. Vectors like $\underline{Q}_6$ and $\underline{Q}_7$, for which $\underline{Q}.\underline{c}$ equals 2π (hence $\ell = 1$), all end in a plane which is parallel to the $\ell = 0$ plane and distant $2\pi/c$ from it. Similarly the vectors $\underline{Q}_8$ and $\underline{Q}_9$ define another plane with $\ell = 2$. For $\underline{Q}_1$, $\underline{Q}_2$ and $\underline{Q}_3$, the dot product $\underline{c}.\underline{Q}$ equals -2π. Hence these vectors define a plane which is distant $2\pi/c$ from the $\ell = 0$ plane, but in the opposite direction to the $\ell = 1$ plane; this plane has $\ell = -1$. Since $S(\underline{Q})$ is non-zero only on these planes, it follows from equation VIII.4, that $I(\underline{Q})$ too is non-zero only on them.

Figure VIII.5 shows that the diffraction pattern is confined to a series of lines, known as "layer lines"; these lines are formed when the planes, on which $I(\underline{Q})$ is non-zero, intersect the surface of the Ewald sphere. In the figure a beam of X-rays is incident at right angles to the c-axis of a one-dimensional crystal. Here O, the origin of real space, is defined by the intersection of the beam with the c-axis. By definition O', the origin of Q-space, is distant $2\pi/\lambda$ from O along the direction of the undeflected beam. $I(\underline{Q})$ is non-zero only on a set of planes, spaced $2\pi/c$ apart in Q-space, which are perpendicular to the c-axis direction. As shown in Fig. VIII.5, the planes intersect the Ewald sphere along a series of lines. These lines, referred to as "layer lines", can be specified, or "indexed", by the value of ℓ. By definition the plane containing O' has a zero value of ℓ. The corresponding layer line, with $\ell = 0$, is often called the "equator"; the line on the diffraction pattern which is parallel to the c-axis, and therefore perpendicular to the equator, is called the "meridian".

In Fig. VIII.5 the ζ-axis of Q-space is defined to be parallel to the c-axis of real space; naturally it passes through O' the origin of Q-space. The planes on which $I(\underline{Q})$ is non-zero are perpendicular to the c-axis and, therefore, to the ζ-axis. The ℓ th plane of Q-space crosses the c-axis at a distance

$$\zeta = 2\pi\ell/c \qquad \text{(VIII.6)}$$

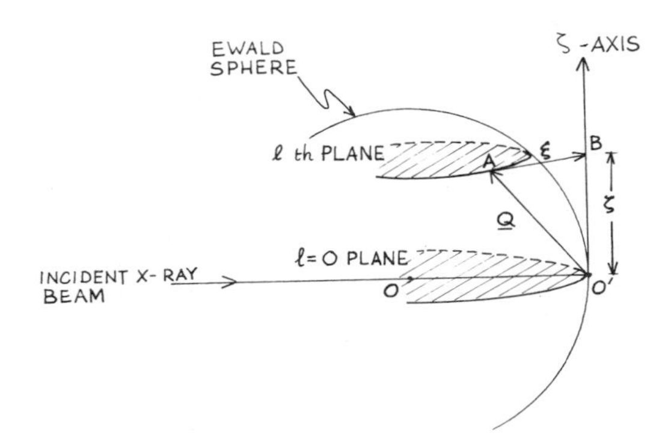

Fig. VIII.6. Definition of ζ and ξ.

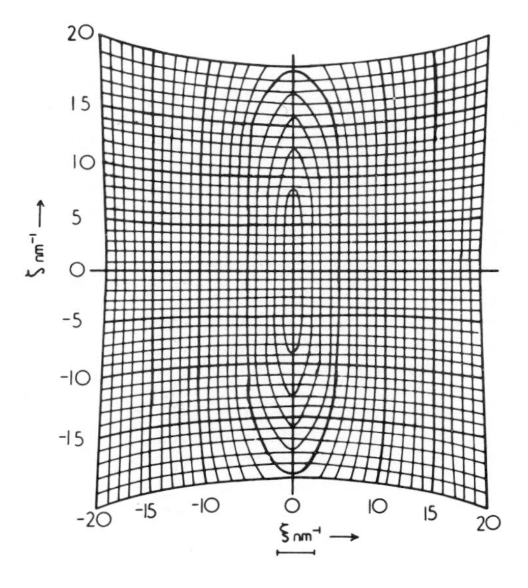

Fig. VIII.7. Bernal chart for the diffraction pattern from a one-dimensional crystal where the incident X-ray beam is perpendicular to the c-axis recorded on a flat photographic film with a specimen-to-film distance of 10 cm.

TABLE VIII.1. Relationship between film (x,y) and Q-space (ξ,ζ) coordinates for a flat film perpendicular to the incident X-ray beam.

(The axis of the one-dimensional crystal is perpendicular to the X-ray beam whose wavelength is λ)

$$x = R \tan \cos^{-1} \{\gamma/2[1 - (\zeta\lambda/2\pi)^2]^{\frac{1}{2}}\}$$

$$y = 2R(\zeta\lambda/2\pi)/\gamma$$

$$\gamma = 2 - (\zeta\lambda/2\pi)^2 - (\xi\lambda/2\pi)^2$$

from O'. Except for the trivial case when ℓ equals zero, the ζ-axis does not touch the surface of the Ewald sphere. However, if c is sufficiently large, ζ will be very small for the first few planes of non-zero I(Q). Consequently these intersect such a limited region of the surface of the Ewald sphere, around O', that it is effectively a plane which is perpendicular to the incident beam direction. Then ζ effectively lies on the surface of the sphere for small values of ℓ. As a result the meridian of the diffraction pattern corresponds to the ζ-axis of Q-space, for layer lines with low ℓ values, provided c is sufficiently large. The distances between layer lines can be measured on the diffraction pattern and converted into Q-space using equation IV.2; they can then be equated with ζ of equation VIII.6 in order to determine c.

A device called a Bernal chart can be used to measure ζ from the diffraction pattern even when c is small - equation VIII.6 can then be used to calculate c from ζ. Figure VIII.6 shows a vector Q which terminates at A - the intersection of the Ewald sphere with a point on the ℓ th plane on which I(Q) is non-zero. Thus the intensity at A gives rise to a point on the diffraction pattern. If the pattern is recorded on a flat film, the distance of this point from the centre of the pattern (the intersection of the equator and meridian) is related to Q by equation IV.2. Now Q has components ζ and ξ which are, respectively, parallel and perpendicular to the ζ-axis. (ξ is the distance of A from B - B is the point where the ℓ th plane intersects the ζ-axis, as shown in Fig. VIII.6.) Similarly the corresponding point on the film has equatorial and meridional components x and y. (Note that here x is a component of x and not its modulus - in contrast to the usage in equation IV.2.) Figure IV.5 showed how the diffraction pattern arose from projecting Q-space on a flat film. Considerable trigonometric manipulation is required to relate the components in Q-space to those on the film - see Volume II of the "International Tables for X-Ray Crystallography" (BIBLIOGRAPHY Section 2). The results are given in Table VIII.1.

Figure VIII.7 shows a Bernal chart; it is essentially a two-dimensional ruler for measuring ξ and ζ directly from the diffraction pattern. For a given value of R, the relationships of Table VIII.1 can be used to plot lines which pass through equal values of ζ and ξ on the film. R, the distance between the specimen and the film, can be readily measured during an experiment. In practice the measurements of ζ and ξ are best made on a photographic enlargement of the diffraction pattern. (But measurements of intensity must always be made on the original pattern.) The scale of the enlargement can be measured and the Bernal chart enlarged by the same factor. Then the chart is laid on the pattern and the value of ζ corresponding to the layer line separation is read directly. The spacing c

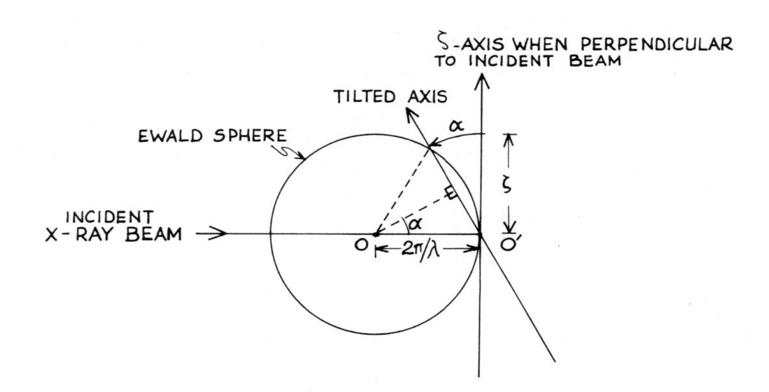

Fig. VIII.8. Section of the Ewald sphere showing the effect of tilting the ζ-axis.

can be calculated from ζ using equation VIII.6.

The intensity distribution along the ζ-axis can be recorded by tilting the one-dimensional crystal so that its c-axis is no longer perpendicular to the beam. Figure VIII.8 is a section through the Ewald sphere. It shows that the ζ-axis will touch the Ewald sphere on the ℓ th layer line if it is tilted through an appropriate angle, α, where

$$\sin \alpha = (\zeta/2)/(2\pi/\lambda) \qquad \text{(VIII.7)}$$

From equations VIII.6 and VIII.7:

$$\alpha = \sin^{-1} (\ell\lambda/2c) \qquad \text{(VIII.8)}$$

A separate tilt angle is required for each of the planes on which $I(\underline{Q})$ is non-zero, according to equation VIII.8, since ℓ has a different value for each plane. How can the ζ-axis be tilted? It follows from the rotation properties of the Fourier transform, in Section II.4, that a tilt in real space is accompanied by an equal tilt, in the same direction, about a parallel axis in Q-space. Thus, in order to tilt the ζ-axis as required, the crystal c-axis is tilted so that it departs, by an angle α, from the perpendicular to the incident beam.

All these measurements rely on two presumptions. One: that the specimen is known to be a one-dimensional crystal. Two: that the specimen can be set up so that its c-axis is perpendicular to the incident X-ray beam. If the pattern consists of a series of layer lines of comparable sharpness, then the specimen can be assumed to be a one-dimensional crystal. According to Section III.8, $I(\underline{Q})$ is identical to $I(-\underline{Q})$ - provided that there are no atoms present with an absorption edge close to the incident wavelength ie. if anomalous scattering is negligible. Thus when the ζ-axis, and hence the c-axis, is perpendicular to the X-ray beam, the intensity distribution will conform to the condition that $I(\underline{Q})$ is identical to $I(-\underline{Q})$. The tilt of the specimen can be adjusted, until this condition is met, to make the c-axis perpendicular to the incident beam.

VIII.4. Intensity of the layer lines

The layer lines in Fig. VIII.6 need not all have the same intensity. According to equation VIII.4, a non-zero value of $S(\underline{Q})$ only determines that the lines are allowed. The intensity of a line is given by $F(\underline{Q})F^*(\underline{Q})$ amplified by the value of $S(\underline{Q})$ - which is the same for every layer line but zero elsewhere. Since $F(\underline{Q})$ is the Fourier transform of the electron density in a layer of the one-dimensional crystal, it is the structure of one of these layers, shown in Fig. VIII.2, which determines the intensity along a layer line.

The equator provides information about the structure of the one-dimensional crystal projected on to the plane of a layer ie. perpendicular to its c-axis. Section IV.4 indicated that a plane section of Q-space, which passes through O', provides information about the scattering specimen projected on to a parallel plane in real space. This property follows directly from the definition of the Fourier transform in equation II.6 and leads to the special case described here. As in Section II.4, it is not intended to prove the general properties of the Fourier transform here - the reader is referred to books by Bracewell and Champeney (see BIBLIOGRAPHY Section 4).

Similarly the ζ-axis provides information about the structure of the crystal projected on to the c-axis. This result follows from exactly the same property of the Fourier transform. Now a line in Q-space, which passes through O', provides information about the scattering specimen projected on to a parallel line in real space. Once again this property follows directly from the definition of the Fourier transform in equation II.6.

An expression will now be derived for the Fourier transform along the ζ-axis - note that the corresponding intensity distribution only corresponds to the meridian of the diffraction pattern if c is very large and, even then, only at low ℓ values. We have seen that the Fourier transform of a one-dimensional crystal is given by $F(\underline{Q})\psi(\underline{Q})$ where $\psi(\underline{Q})$ simply leads to the intensity, and hence the transform, $F(\underline{Q})$, of a layer in the one-dimensional crystal being amplified along certain planes of Q-space but obliterated elsewhere. Thus, in order to calculate the Fourier transform of a one-dimensional crystal, we need only calculate the transform of a single layer - and the calculation can be confined to the planes in Q-space which are specified by integral values of ℓ.

Since we are concerned with the transform along the ζ-axis, we need to calculate the Fourier transform of the electron density, of a single layer of the one-dimensional crystal, projected on to the c-axis. The projected electron density in a single layer is denoted by $\rho(Z)$; where a displacement along the c-axis is denoted by

$$r = Zc \tag{VIII.9}$$

Here Z is a fractional translation since c, as in Fig. VIII.1, is the distance between the crystal layers; if Z were greater than unity we would be considering the structure of the next layer. Since we are only calculating the Fourier transform along the ζ-axis, which is parallel to c, the dot product, which arises in the definition of the Fourier transform, becomes

$$\underline{r}.\underline{Q} = rQ = Zc\zeta = 2\pi\ell Z \tag{VIII.10}$$

from equations VIII.6 and VIII.9. Equation VIII.10 gives the values of $\underline{r}.\underline{Q}$ along the ζ-axis where the Fourier transform of the one-dimensional crystal is non-zero. Thus, from equations II.6 and VIII.10, the Fourier transform of a single layer projected on to the c-axis direction, when it forms part of a one-dimensional crystal, is given by

$$F(\ell) = \int_0^1 \rho(Z) \exp(2\pi i \ell Z)\, dZ \qquad \text{(VIII.11)}$$

Note that points along the ζ-axis, where $F(\underline{Q})$ is non-zero, are specified by values of ℓ and so the transform at these points can be written as $F(\ell)$.

VIII.5. Deduction of electron density

In the diffraction patterns from many real examples of one-dimensional crystals, $I(\underline{Q})$ is reasonably intense only on the ζ-axis and on the plane with $\ell = 0$. If c is large, the ζ-axis appears on the meridian. Since $I(\underline{Q})$ decreases rapidly, in intensity, off the ζ-axis, the meridian then appears as a row of spots. These spots will have spacings which correspond to small values of ζ and are often referred to as "low-angle meridional" patterns.

The electron density of a layer projected on to the c-axis direction, $\rho(Z)$, can be deduced from a low-angle meridional pattern - provided that the phase problem, of Section III.2, is soluble. Fourier inversion of equation VIII.11 yields, according to Section II.4:

$$\rho(Z) = \sum_{\ell} F(\ell) \exp(-2\pi i \ell Z) \qquad \text{(VIII.12)}$$

Here a summation, rather than an integral, is appropriate because $F(\ell)$ is confined to a series of discrete points. In practice the intensity, $I(\ell)$, is measured at these points - the real and imaginary parts of $F(\ell)$ then have to be recovered from its modulus, $I^{\frac{1}{2}}(\ell)$, by one of the methods of Section III.4. Recovery of these parts constitutes a solution to the phase problem. Note that, since $F(\ell)$ is not a smoothly varying function, the so-called "direct methods" which are appropriate in this case are far more subtle than the simple example given in Section III.4.

Omission of F(O) does not distort the form of $\rho(Z)$ calculated using equation VIII.12. Thus it is of no concern that the intensity of the undeflected beam is not measured. Equation VIII.12 can be rearranged to give

$$\rho(Z) = F(0) + \sum_{\ell \neq 0} F(\ell) \exp(-2\pi i \ell Z)$$

The result shows that F(O) acts only as a constant background which does not affect the form of $\rho(Z)$ - its omission has the effect of removing a constant background from the calculated electron density. In the general case of Section III.7 there was the possibility of scattered intensity very close to the undeflected beam which carried information about fluctuations in the structure. Removal of the undeflected beam led to removal of the neighbouring scatter, at very low Q values, with the consequent loss of structural information. Here the information carried by the scattered X-rays is confined to the layer lines and is easily separated.

The calculated electron density distribution, $\rho(Z)$, can be distorted by loss of diffraction data at high Q values. Strictly equation VIII.12 should include all ℓ values between minus and plus infinity. Loss of higher ℓ values corresponds to limiting the extent of Q-space which is used. According to Section V.4, the electron density can then only be calculated at low resolution. More seriously, spurious peaks can appear in $\rho(Z)$ - as shown in Section III.7.

Therefore it is not necessarily valid to attribute peaks in the calculated $\rho(Z)$ to features in the scattering specimen. Suppose that a peak appears in $\rho(Z)$, like that in Fig. III.7 - does it represent a true structural feature? This model for $\rho(Z)$ can be used to compute the expected meridional intensity distribution, $I(\ell)$, from equations VIII.11 and II.5, to compare with that observed. Other models, in which the peak is smoothed out, could also be used to calculate values of $I(\ell)$ and the levels of agreement with experimental data compared to the agreement for the peaky model. There would only be evidence for the peak representing a real structural feature if the level of agreement for the peaky model were significantly better than for the others. A suitable statistical test, to determine whether such a difference is significant, is described in Volume IV of the "International Tables for X-Ray Crystallography" (see BIBLIOGRAPHY Section 2).

VIII.6. Bragg's law

Bragg's law is often used to explain X-ray diffraction patterns from crystals. Consider the X-rays scattered so that $\underline{Q}$ is parallel to the vector $\underline{c}$ which separates the points in Fig. VIII.2. (In this experiment the incident beam would not be perpendicular to the c-axis.) This row of points is the one-dimensional lattice of the crystal. For these X-rays the dot product $\underline{c}.\underline{Q}$ becomes Qc and equation VIII.5 becomes

$$Q = 2\pi\ell/c \tag{VIII.13}$$

Comparison of equations II.4 and VIII.13 yields

$$(4\pi/\lambda)\sin(\phi/2) = 2\pi\ell/c \tag{VIII.14}$$

where ϕ is the angle through which the X-rays are scattered. If the angle θ is defined as $\phi/2$, equation VIII.14 becomes

$$2c \sin\theta = \ell\lambda$$

which is the form of Bragg's law which conventionally appears in books on X-ray crystallography.

Thus Bragg's law arises automatically from the Fourier transform of a crystal. Why adopt this apparently more complicated approach instead of deriving Bragg's law by the simple method given in elementary physics textbooks? The reason is that Bragg's law only describes when $I(\underline{Q})$ might be expected to be non-zero; it provides no further information about $I(\underline{Q})$ and gives no clue as to how a structure might be determined. Furthermore the use of Bragg's law to explain interference effects can be very misleading because it applies only to crystals. The approach adopted in this book shows how structure can be determined (as described in Chapter III) and is perfectly general - it applies to all states of matter.

VIII.7. Disorder

No crystal is perfect. In Fig. VIII.1 the bodies cannot all be expected to lie exactly on the planes spaced a distance c apart. For one reason or another a body might have a displacement, r, along the c-axis direction - above or below its plane. Figure VIII.9 shows the resulting, one-dimensional crystal. Let the root-mean-square displacement be σ. If the distribution is Gaussian, σ represents the standard deviation of the displaced distances whose distribution is then given by

$$D(r) = \{1/\sigma(2\pi)^{\frac{1}{2}}\}\exp(-r^2/2\sigma^2) \tag{VIII.15}$$

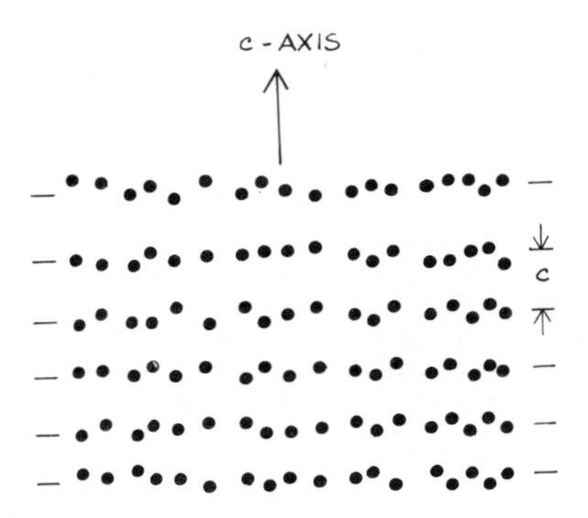

Fig. VIII.9. Imperfect one-dimensional crystal.

The lattice of an imperfect crystal is generated by convolution of the perfect one-dimensional lattice in Fig. VIII.2 with D(r). In this figure the row of points now represents the mean positions of the bodies along the c-axis direction - the convolution generates the distributions of positions about these means. Following the method of Section VIII.2, it is necessary to calculate the Fourier transform of this imperfect lattice in order to derive the interference function for an imperfect, one-dimensional crystal. The Fourier transform of the imperfect lattice is obtained by multiplying the transform of the perfect lattice by the transform of D(r). Fourier transformation of D(r) yields

$$\tilde{D}(Q) = \exp\ (-\ Q^2\sigma^2/2) \qquad \text{(VIII.16)}$$

Equation VIII.16 is derived from equation VIII.15 using a general expression for the Fourier transform of an exponential function given by Champeney (see BIBLIOGRAPHY Section 4). The interference function for a perfect one-dimensional crystal was obtained, in Section VIII.2, by multiplying the Fourier transform of the perfect lattice by its complex conjugate. For an imperfect one-dimensional crystal we have to multiply the Fourier transform of the imperfect lattice by its complex conjugate. Since $\tilde{D}(Q)$ is real, the result is simply the interference function for a perfect lattice multiplied by the square of $\tilde{D}(Q)$.

The effect of this disorder is to decrease the intensity of the layer lines without altering their sharpness. It is, therefore, very different from the effect of disturbing the arrangement of bodies, in the c-axis direction, until it resembles an amorphous solid; then the peaks in S(Q) would increase in width with increasing Q - see Fig. VII.6. When σ is zero, the value of $\tilde{D}(Q)$ is unity and the intensity is unchanged; this result is physically reasonable because a zero value of σ corresponds to a perfect one-dimensional crystal. As σ departs from zero, the value of $\tilde{D}(Q)$ decreases very rapidly, because it is an exponential function, so that the intensity of the layer lines is greatly reduced. Multiplication by the square of $\tilde{D}(Q)$ can never increase the value of the interference function of the perfect crystal and so the peak width in Fig. VIII.3, and hence the sharpness of the layer lines, is unchanged.

Diffuse intensity appears between the layer lines of the diffraction pattern from an imperfect one-dimensional crystal. When the intensity of a layer line decreases, the X-rays scattered in this direction transmit less energy. The energy supplied to the specimen, by the incident beam, is unchanged by the nature of the specimen. No more energy is absorbed: X-rays are absorbed by matter when electrons are promoted to higher energy levels - so absorption by a given number of identical bodies is largely independent of their positions. If the principle of conservation of energy is not to be violated, the energy lost from the layer lines must be used to scatter X-rays in other directions - which leads to the appearance of weak, diffuse intensity between them. A rigorous account of diffuse intensity is given by Guinier, for a three-dimensional crystal; James gives far more detail (see BIBLIOGRAPHY Section 2).

We have seen that the disorder described here is distinct from the structural irregularity of an amorphous solid; it can arise in two ways - thermal vibration and random displacement of bodies about lattice positions. In the paracrystal theory of disordered systems, the random displacement about lattice points is called "disorder of the first kind". Thermal vibration and this structural displacement can only be distinguished experimentally by measuring diffraction patterns at a variety of temperatures. Extrapolation to absolute zero temperature, at which there is no thermal vibration, then gives the structural displacement alone.

VIII.8. Mosaic spread

Imperfection can arise in another way - the c-axis of Fig. VIII.1 may not point in exactly the same direction all over the crystal. This kind of imperfection is a one-dimensional example of mosaic spread - although in the one-dimensional case it is usually referred to as "imperfect orientation". We can consider the mosaic one-dimensional crystal to consist of an assembly of perfect one-dimensional crystallites, each with a different c-axis direction.

The diffraction pattern of a mosaic crystal is formed by adding the intensity contribution from each ie. interference effects between X-rays scattered by different crystallites are negligible. *A priori* we might have considered a need for both intra- and inter-crystallite interference functions; $S(\underline{Q})$ of equation VIII.4 is the former. However, the crystallites are so large, compared with the distances between their constituent scattering bodies, that inter-crystallite interference effects will be appreciable only at very low Q values - from the reasoning of Section V.4. There is no reason to suppose that the crystallites will be regularly arranged in the crystal. By analogy with the interference function of a liquid, in Fig. VII.5 and VII.6, the inter-crystal interference function will have settled to a value of unity for the region of Q-space in which intra-crystallite interference effects are observed. Since inter-crystallite interference is negligible, the diffraction pattern is formed by adding the intensity contributions of the crystallites - as in the ideal gas theory of Section VI.2.

In the mosaic crystal the c-axes of the crystallites will be distributed about some preferred direction; consequently a corresponding distribution of vectors occurs in Q-space. For a perfect one-dimensional crystal the distribution of tilts is infinitely narrow ie. there is a unique c-axis as in Fig. VIII.1. If the c-axis of a crystallite is tilted it follows, from Section II.4, that an equal tilt, in the same direction, about a parallel axis through O', must be applied to its Fourier transform. Since there is a distribution of tilts, a point in Q-space is spread over a region. Although the c-axis tilts are applied to different regions of real space, all tilts in Q-space are referred to the same origin, O'. The reason is that, according to the definition of equation II.6, a shift of real space position, $\underline{r}$, only affects the phase of the Fourier transform, $F(\underline{Q})$, and not the point in Q-space at which $I(\underline{Q})$, given by equation II.5, has a particular value.

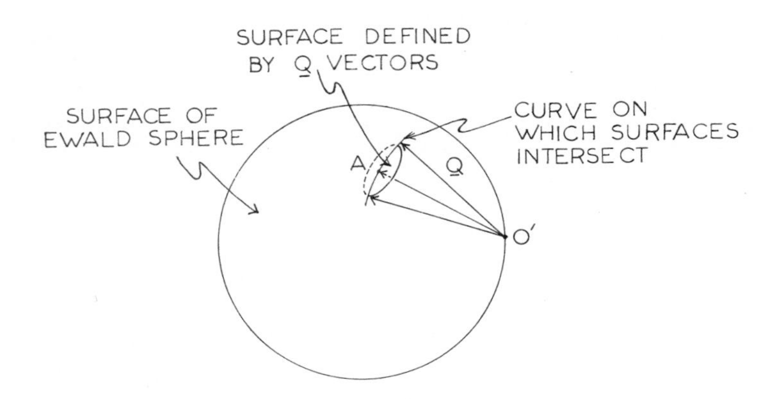

Fig. VIII.10. Mosaic spread distributes the intensity at the point A of Fig. VIII.6 over a surface in Q-space; this surface intersects the Ewald sphere along an arc.

Suppose a vector $\underline{Q}$ would have terminated at a point A, in Q-space, for a perfect one-dimensional crystal; it terminates on a portion of a spherical surface, in Q-space, for a mosaic one-dimensional crystal. In the perfect crystal $\underline{Q}$ had a unique direction; now this direction is not unique but is merely preferred. There is a family of $\underline{Q}$ vectors, all with the same modulus Q, but with a distribution of tilts about the preferred direction, in the mosaic case. The intensity, $I(\underline{Q})$, of the point in the perfect crystal will not be evenly distributed over this surface but will have its maximum value at the original point which now corresponds to the preferred, rather than the unique, direction of the c-axis.

Points on the diffraction pattern of a perfect one-dimensional crystal are drawn into arcs on the diffraction pattern of a mosaic one-dimensional crystal. Fig. VIII.10 shows a point A on the surface of the Ewald sphere - as in Fig. VIII.6. Now there is a family of vectors, all of modulus Q, terminating around A. These vectors describe a portion of the surface of a sphere of radius Q centred at the origin of Q-space, O'. A lies both on this portion of spherical surface and on the surface of the Ewald sphere. Figure VIII.10 shows that the two surfaces intersect along an arc through A. On a diffraction pattern an arc will appear instead of a point. Furthermore, only a part of $I(\underline{Q})$ will be recorded on the pattern because only part of the surface over which it is distributed intersects the surface of the Ewald sphere.

VIII.9. Lorentz correction

The need for the so-called "Lorentz correction" arises because different points in Q-space have different opportunities to touch the surface of the Ewald sphere. Figure VIII.5 provides an example. The intensity distribution along ζ cannot be recorded, except on the $\ell = 0$ plane, unless the specimen is tilted. Even then the required intensity information is only given on the layer line whose value of ℓ is related to the tilt angle, α, by equation VIII.8.

Mosaic spread, described in Section VIII.8, causes a further complication for real systems. This effect causes a point in Q-space to be distributed over part

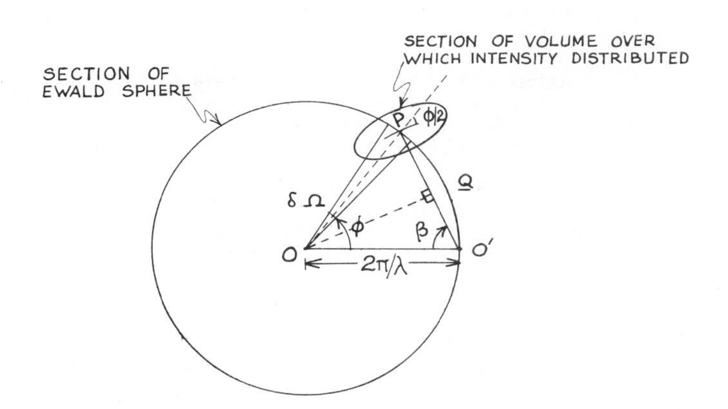

Fig. VIII.11. Figure for the derivation of the Lorentz factor.

of the surface of a sphere, of radius Q, as shown in Fig. VIII.10. For a given specimen this surface area will increase in proportion to Q^2. But only an arc, whose length will be proportional to Q, intersects the surface of the Ewald sphere and can appear on the diffraction pattern. Thus the fraction of $I(\underline{Q})$ which is recorded decreases as Q increases. Note that this spread has another effect. Once again Fig. VIII.5 provides an example. Some points along the ζ-axis may be spread out so much that they actually touch the surface of the Ewald sphere. Then, although the theory for a perfect one-dimensional crystal predicts that they cannot appear on the diffraction pattern, they may appear in practice. However only a small fraction of the total $I(\underline{Q})$ will be recorded.

Rotation of Q-space about an axis, through O', perpendicular to the plane of Fig. VIII.8, might appear to simplify the analysis of the intensity distribution along the meridian in favourable cases. According to Section II.4, this rotation could be achieved by an equal rotation of real space about a parallel axis through O. In many real examples of one-dimensional crystals, the intensity distribution along the ζ-axis is much more intense than in the surrounding region of Q-space. The rotation would ensure that all of the intensity of the points along the ζ-axis would pass through the surface of the Ewald sphere. All possible values of the tilt angle, α, in Fig. VIII.8, would be used while recording a single diffraction pattern. Alternatively, Q-space could be oscillated, backwards and forwards, through the surface of the Ewald sphere, over a sufficient angular range, to ensure that all of the intensity over the points would pass through. In each case the only limitation of the points that could be recorded is that their Q value does not exceed the maximum value which can be recorded for radiation of the wavelength being used - see Section III.6.

An analysis of this rotation shows how the Lorentz factor arises and allows its form to be calculated. Points are spread over part of the surface of a sphere, in Q-space, by mosaic spread, and the planes on which the intensity is non-zero have a finite width - as shown in Fig. VIII.3. Thus, although theory yields the intensity at a point in Q-space, this intensity is actually spread over a volume of Q-space - where the intensity distribution per unit volume is $q(\underline{Q})$, say. Figure VIII.11 shows a section through the Ewald sphere. In this figure the volume of Q-space, over which the intensity associated with the point P is distributed, intersects the surface over an area which subtends a solid angle of Ω at O, the origin of real space.

Suppose the volume of Q-space is rotated through the surface of the Ewald sphere about an axis, passing through O', which is perpendicular to the plane of Fig. VIII.11. Consider an element, $\delta\Omega$, of the solid angle Ω. It corresponds to an area on the surface of the Ewald sphere of

$$(2\pi/\lambda)^2 \ \delta\Omega$$

For an increase, $\delta\beta$, in the angle β of Fig. VIII.11, an arc of length

$$Q\delta\beta$$

passes through the surface of the sphere. The volume when an elemental area passes through the surface is given by the product of this area and the component of this arc which is perpendicular to it. From the figure this component is

$$Q\delta\beta \cos(\phi/2)$$

Thus the volume element in Q-space which passes through the surface of the Ewald sphere is given by

$$\delta V = (2\pi/\lambda)^2 \ \delta\Omega \ Q\delta\beta \cos(\phi/2) \tag{VIII.17}$$

From equations II.4 (which relates Q to ϕ) and VIII.17

$$\delta V = (2\pi/\lambda)^3 \sin\phi \ \delta\beta \ \delta\Omega \tag{VIII.18}$$

The rotation allows the whole volume of Q-space, over which the intensity of the point is spread, to pass through the surface of the Ewald sphere. Thus the intensity recorded on the diffraction pattern is given by

$$I(\underline{Q}) = \int_\beta \int_\Omega q(\underline{Q}) \ d\Omega \ d\beta \tag{VIII.19}$$

where the subscripts β and Ω denote integration over the whole rotation as well as over the whole solid angle subtended by the surface through which the volume passes. From equations VIII.18 and VIII.19:

$$\begin{aligned} I(\underline{Q}) &= (\lambda/2\pi)^3 \ (1/\sin\phi) \int_\beta \int_\Omega q(\underline{Q}) \ dV \\ &= (\lambda/2\pi)^3 \ (1/\sin\phi) \ F(\underline{Q})F^*(\underline{Q}) \end{aligned} \tag{VIII.20}$$

since the Fourier transform, $F(\underline{Q})$, multiplied by its complex conjugate, gives the total intensity over this volume of Q-space.

In an experiment we may wish to deduce the modulus of $F(\underline{Q})$ from the intensity, $I(\underline{Q})$, of a spot whose centre is given by the vector $\underline{Q}$. It is then necessary to multiply this experimental intensity by $\sin\phi$, the Lorentz factor, before this modulus can be equated with $I^{\frac{1}{2}}(\underline{Q})$. The $(\lambda/2\pi)$ term can be ignored as it is a constant for a given X-ray wavelength. The Lorentz factor can be calculated from Q by using equation II.4 which relates ϕ to Q.

VIII.10. Summary

A one-dimensional crystal has a regularly repeating structure in a single direction (its c-axis). Its Fourier transform and, hence, the intensity of the X-rays which it scatters are non-zero only on a set of planes in Q-space. These planes intersect the surface of the Ewald sphere along lines - when these lines appear on the diffraction pattern, they are called layer lines. The layer line which includes the undeflected incident beam is called the equator and is assigned a zero value of an integer ℓ; the distribution of scattered intensity along the equator provides information about the structure of the one-dimensional crystal projected on to a plane perpendicular to the c-axis. Successive layer lines, and the planes which gave rise to them are assigned ℓ values of 1, 2, 3 - - - on one side of the equator and -1, -2, -3 - - - on the other. Layer lines on the diffraction pattern are perpendicular to the c-axis of the one-dimensional crystal and are $2\pi/c$ apart when converted into Q-space, where c is the repeat distance of the structure along the c-axis.

The line on the diffraction pattern which is parallel to the c-axis is called the meridian. If the c-axis is suitably tilted, the structure of the one-dimensional crystal, in projection on to the c-axis, can be deduced from the meridional intensity distributions of a series of patterns. Tilting may not be necessary for specimens with large c values if only low resolution information about the structure is required. A Lorentz correction has to be applied to the measured intensity distribution - in part because a point in Q-space is drawn into an arc on the diffraction pattern by mosaic spread. In a real one-dimensional crystal the scattering bodies are usually distributed about the positions of the ideal case. This distribution leads to a diminution of the intensity of the layer lines and the appearance of weak diffuse intensity between them - the pattern does not resemble that of an amorphous solid if the displacements are small and distributed about the ideal positions.

CHAPTER IX
Helices and Liquid Crystals

IX.1. Introduction

This chapter is concerned with diffraction by systems which provide real examples of one-dimensional crystals. Many polymer molecules are one-dimensional crystals themselves. The reason is that, under suitable conditions, many of these molecules are helical - and a helix has the repetitive structure in a single direction which characterises a one-dimensional crystal. Several biological structures are helical - some, such as the "thin filaments" of muscle, consist of globular macromolecules arranged on a helix. Before the diffraction properties of any of these helices can be discussed, it will first be necessary to consider the Fourier transforms of two geometric entities - the continuous helix and the discontinuous helix.

Smectic liquid crystals provide a further example of one-dimensional crystals - although some types of smectic have a degree of order in three dimensions. Although the nematic liquid crystals, described in Section IX.5, are not at all crystalline, even in one dimension, reasons will be given for considering their diffraction properties in this chapter. There is also a practical reason for their inclusion after helices. A discussion of diffraction by helical molecules will be inseparable from consideration of polymers. An important application of X-ray diffraction in polymer science is the determination of any preferred directions in which polymer chains are oriented in a sample. The least ordered arrangement of molecules in an oriented polymer sample is the same as that in a nematic liquid crystal.

IX.2. Continuous helix

Figure IX.1 shows that the familiar corkscrew shape of a continuous helix can be generated by the convolution of a single turn with a lattice of points, spaced p apart, where p is the helix pitch; comparison with Fig. VIII.2 shows that the helix is a one-dimensional crystal. Convolution of a function with a set of points will be considered frequently in this chapter - the operation is represented pictorially in Fig. II.5. Because the helix is a one-dimensional crystal, its Fourier transform is non-zero only on a set of planes, spaced $2\pi/p$ apart in Q-space, which are perpendicular to the direction of its axis - provided the helix is of effectively infinite extent. Each plane is specified by the value of the integer ℓ and gives rise to a layer line on the diffraction pattern as described in Section VIII.3; the undeflected X-ray beam defines the centre of the diffraction pattern

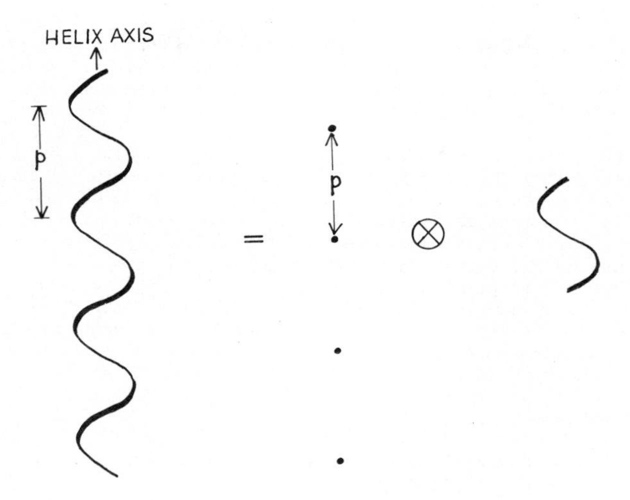

Fig. IX.1. Generation of a continuous helix by convolution.

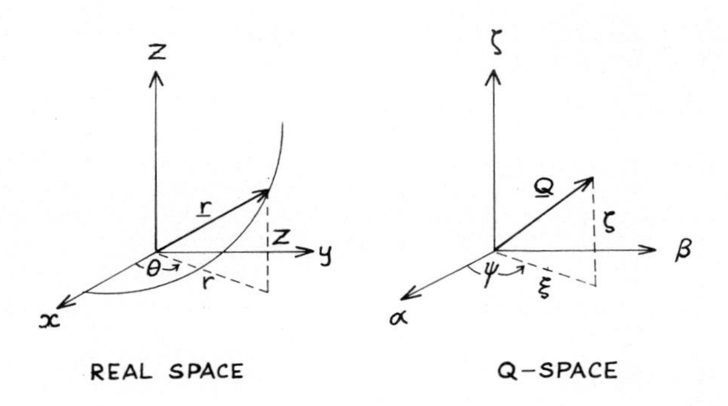

Fig. IX.2. Coordinate systems in real space and Q-space.

and lies on the layer line with $\ell = 0$.

According to Section VIII.4, the Fourier transform of the helix is obtained by calculating the transform of a single turn on each of these planes. Figure IX.2 defines the Cartesian (x,y,z) and cylindrical polar (r,θ,z) coordinates which can be used to describe a point on the helix which is defined by the vector $\underline{r}$. In this section the helix will be considered to have infinitesimal thickness - like a corkscrew made of infinitely thin wire. Note that here r is the projection of $\underline{r}$ on to a plane perpendicular to the helix axis (the z-axis) and not its modulus; also the symbols x and y are unrelated to their usage in Sections IV.3 and VIII.3. The value of r is constant for a given helix which is completely specified by the values of r and p. Figure IX.2 also shows the corresponding Cartesian (α,β,ζ) and cylindrical polar (ξ,ψ,ζ) coordinates which will be used to describe a point in Q-space defined by the vector $\underline{Q}$. The z-axis of real space and the ζ-axis of Q-space are defined to be parallel as in Fig. VIII.6; the direction of an X-ray beam, which is incident at right angles to the helix axis, could be taken to define the directions of the x-axis and the α-axis. However all that is necessary is for these axes to be defined to be parallel. Once again α and β are unrelated to their usage in Sections VIII.3 and VIII.9. Readers who do not wish to follow the detailed derivation of the Fourier transform of the continuous helix should skip the next four paragraphs.

In Fig. IX.2 the components of $\underline{r}$ are the Cartesian coordinates of the point which it defines; thus it can be represented by

$$\underline{r} = \begin{bmatrix} x \\ y \\ z \end{bmatrix} = \begin{bmatrix} r \cos 2\pi z/p \\ r \sin 2\pi z/p \\ z \end{bmatrix} \qquad \text{(IX.1)}$$

since the helix turns through an angle of 2π radians within a single pitch. Similarly

$$\underline{Q} = \begin{bmatrix} \alpha \\ \beta \\ \zeta \end{bmatrix} \qquad \text{(IX.2)}$$

Since, by definition, the α-axis is parallel to the x-axis, the ζ-axis is parallel to the z-axis and, consequently, the β-axis is parallel to the y-axis, it follows, from equations IX.1 and IX.2, that

$$\underline{r}.\underline{Q} = \alpha r \cos (2\pi z/p) + \beta r \sin (2\pi z/p) + \zeta z \qquad \text{(IX.3)}$$

(Remember the comments on the dot product of vectors in Section VIII.3.) The distribution of matter is constant along the helical path but zero elsewhere. Assigning the electron density along this path a value of unity, on some arbitrary scale, the Fourier transform of a single turn of helix becomes

$$F(\alpha,\beta,\zeta) = \int_0^p \exp i\{\alpha r \cos (2\pi z/p) + \beta r \sin (2\pi z/p) + \zeta z\}\, dz \qquad \text{(IX.4)}$$

from equations II.6 and IX.3 - noting that the only real space variable in equation IX.3 is z.

Equation IX.4 will now be transformed into cylindrical polar coordinates. Simple trigonometry yields

$$\xi = (\alpha^2 + \beta^2)^{\frac{1}{2}}$$

$$\cos \psi = \alpha/(\alpha^2 + \beta^2)^{\frac{1}{2}}$$

$$\sin \psi = \beta/(\alpha^2 + \beta^2)^{\frac{1}{2}}$$

from Fig. IX.2. Thus the {bracketed} term in equation IX.4 can be multiplied by unity in the form

$$\xi/(\alpha^2 + \beta^2)^{\frac{1}{2}}$$

to yield

$$\xi r \cos (2\pi z/p - \psi) + z\zeta$$

using the identity

$$\cos (A - B) = \cos A \cos B + \sin A \sin B$$

Therefore equation IX.4 can be rewritten as

$$F(\xi,\psi,\zeta) = \int_0^p \exp i\{\xi r \cos (2\pi z/p - \psi) + \zeta z\}\, dz \qquad \text{(IX.5)}$$

We now find the values of $F(\xi,\psi,\zeta)$ on a set of planes which are spaced $2\pi/p$ apart in Q-space and are perpendicular to the z-axis and, hence, to the ζ-axis ie. we obtain an expression for the Fourier transform of the entire helix. Noting that p is the distance between lattice points in Fig. IX.1 we substitute the expression for ζ for a one-dimensional crystal, from equation VIII.6, into equation IX.5 to obtain

$$F(\xi,\psi,\ell) = \int_0^p \exp i\{\xi r \cos (2\pi z/p - \psi) + 2\pi\ell z/p\}\, dz \qquad \text{(IX.6)}$$

Now that the Fourier transform is non-zero only on a set of planes, whose position in the ζ-axis direction is denoted by ℓ, it can be considered as a function of ξ, ψ and ℓ rather than of ξ, ψ and ζ.

The next stage of the calculation is to evaluate the integral in equation IX.6. Its form can be simplified somewhat by introducing the fractional coordinate Z, of Section VIII.4, where Z equals z/p. Noting that

$$dZ/dz = 1/p$$

equation IX.6 can be rewritten as

$$F(\xi,\psi,\ell) = p\int_0^1 \exp i\{\xi r \cos (2\pi Z - \psi) + 2\pi\ell Z\}\, dZ$$

Neglecting p, which acts as a constant scaling factor for a given helix, and rearranging

$$F(\xi,\psi,\ell) = \int_0^1 \exp i\{\xi r \cos (2\pi Z - \psi)\} \exp i(2\pi\ell Z)\, dZ$$

we now make the substitutions

$$X = \xi r \quad \text{and} \quad \phi = 2\pi Z - \psi$$

from which it follows that

$$d\phi/dZ = 2\pi$$

and $F(\xi,\psi,\ell)$ can be written as

$$(1/2\pi)\ \exp\ (i\ell\psi) \int_{-\psi}^{2\pi-\psi} \exp\ (iX \cos \phi)\ \exp\ (i\ell\phi)\ d\phi$$

When the integral is evaluated (see page 31 of the book by Watson recommended in BIBLIOGRAPHY Section 4) this expression becomes

$$i^{\ell} \exp\ (i\ell\psi)\ J_{\ell}(X)$$

where $J_n(X)$ is the n th order Bessel function of the first kind, defined in equation II.9. This result is usually expressed in the form

$$\exp\ i\ell(\psi + \pi/2)\ J_{\ell}(X)$$

since

$$\exp\ (i\ell\pi/2) = \cos\ (\ell\pi/2) + i \sin\ (\ell\pi/2)$$

which has the same value as i^{ℓ}. (The reason is to force all the phase information into the exponential term.)

Now that the integral in equation IX.6 has been evaluated the Fourier transform of the helix can be written as

$$F(\xi,\psi,\ell) = J_{\ell}(\xi r)\ \exp\ i\ell(\psi + \pi/2) \qquad \text{(IX.7)}$$

Equation IX.7 is the Fourier transform of a continuous helix. From equations II.5 and IX.7, the intensity on the ℓ th plane of Q-space is given by

$$I(\xi,\ell) = J_{\ell}^2(\xi r) \qquad \text{(IX.8)}$$

Thus the intensity of X-rays scattered by a continuous helix is independent of ψ ie. it has cylindrical symmetry.

On each plane of Q-space waves, represented by the Fourier transform of equation IX.7, spread out rather like the ripples on the surface of a pond when a stone is dropped into it. In this case the disturbance emanates from the ζ-axis. The intensity on the ℓ th plane, according to equation IX.8, is given by a squared Bessel function whose form is shown, for several ℓ values, in Fig. IX.3. Note that on the ℓ th plane this intensity is independent of direction, ie. of ψ, and depends only on the radius, r, of the helix and the distance, ξ, from the ζ-axis. Since it follows from equation II.10 that

$$J_{-n}(X) = (-1)^n\ J_n(X)$$

the intensity on the $-\ell$ th plane is the same as on the ℓ th. (This result could have been deduced from the cylindrical symmetry and the general property of diffraction patterns discussed in Section III.8.)

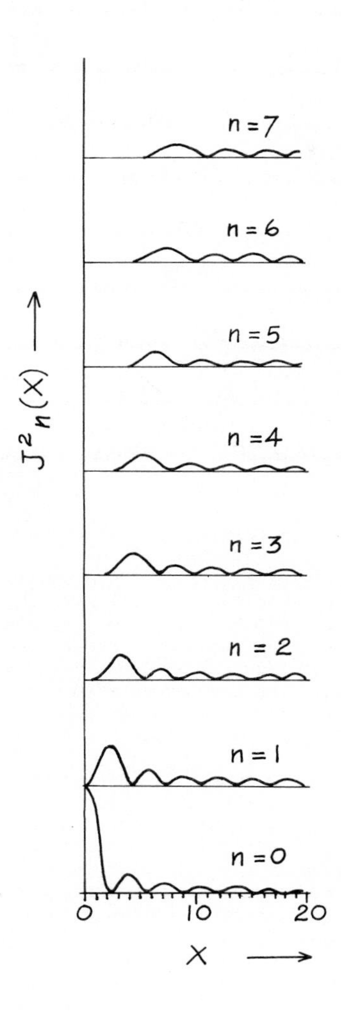

Fig. IX.3. Squared Bessel functions.

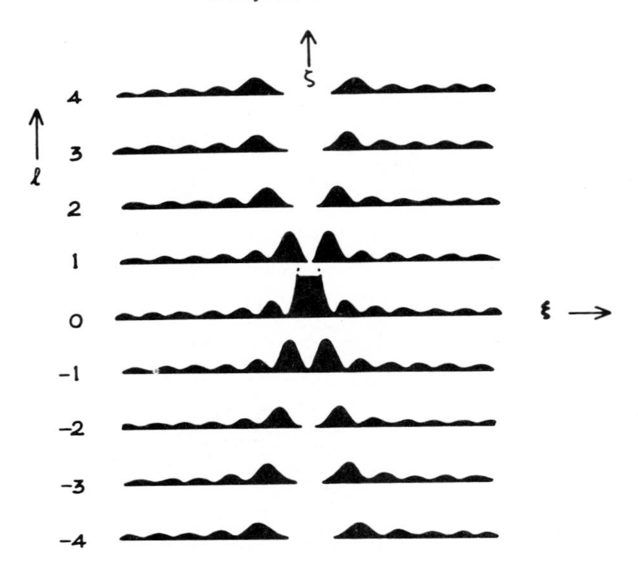

Fig. IX.4. $I(\xi,\ell)$ for a continuous helix.

Figure IX.4 shows the intensity, $I(\xi,\ell)$, in a section of Q-space which contains the ζ-axis. The three-dimensional $I(\xi,\ell)$ can be generated by rotating the figure about the ζ-axis because of its cylindrical symmetry. Figure IX.4 does not quite represent the diffraction pattern to be expected from a continuous helix because the Ewald sphere cannot generally be considered as a plane section of Q-space and because of the distortion introduced when a diffraction pattern is recorded - see Section VIII.3. The figure was produced by plotting the squared Bessel functions of Fig. IX.3 on to a series of layer lines.

Intensity maxima on the layer lines lie on the arms of a diagonal cross and decrease as the magnitude of ℓ increases. Notice that intensity only appears on the ζ-axis when ℓ equals zero. Mathematically this result arises because only the zero order Bessel function in Fig. IX.3 is non-zero when X equals zero. Physically the diffraction pattern must have non-zero intensity at this point because it is defined by the undeflected X-ray beam. The equatorial Fourier transform could have been obtained without recourse to the mathematical derivation of this section. According to Section VIII.4 the equatorial transform is equivalent to the transform of the projection of the helix on to a plane perpendicular to its z-axis. This projection is a circle, ie. an annulus of infinitesimal thickness, so that its Fourier transform, by analogy with the integral of equation II.9, is

$$J_o(Qr)$$

On the equator Q, the modulus of $\underline{Q}$, is identical to ξ so that this result is equivalent to equation IX.7 when ℓ is set equal to zero.

IX.3. Discontinuous helix

A discontinuous helix consists of a series of points which are regularly spaced along a helical path; Fig. IX.5 shows an example where the spacing of points along the helix axis (z-axis) direction is q. For the sake of simplicity we shall initially consider the special case where the helix pitch, p, is an integral multiple of q - a so-called "integral helix". In the figure p/q = 4; thus the example of an integral helix chosen here is a four-fold helix. This figure also shows that a discontinuous helix can be generated by multiplying a continuous helix by a set of planes; the planes are perpendicular to the z-axis, a distance q apart, and represent a three-dimensional function which has a value of zero elsewhere in space. In the figure the helix and the planes are shown in projection so that they appear as a sinusoidal wave and a set of lines, respectively - it is left to the reader's imagination to picture what is happening in three dimensions.

The Fourier transform of a discontinuous helix is given by the convolution of the transform of the continuous helix with the transform of the set of planes. This result follows from the property of the Fourier transform of the product of two functions, described in Section II.4, since the discontinuous helix could be generated by multiplying two functions together. The transform of a set of planes, spaced q apart, which are perpendicular to the z-axis, is a set of points, spaced $2\pi/q$ apart, along the ζ-axis. This result follows from the form of the Fourier transform of a set of points, discussed in Sections VIII.2 and VIII.3 and the inversion property of the Fourier transform, described in Section II.4.

Figure IX.6 represents a section, containing the ζ-axis, through the intensity distribution, $I(\xi,\ell)$, for a discontinuous helix - only the more prominent intensity maxima are shown. In principle this figure is generated by setting down the Fourier transform of a continuous helix at each of the points given by the Fourier transform of the set of planes. Multiplication of the result by its complex conjugate yields Fig. IX.6 - according to equation II.5. In practice the figure is generated by setting down Fig. IX.4 (somewhat simplified here for the sake of clarity) at each of the points, spaced $2\pi/q$ apart, as origin.

Note that Fig. IX.6 does not correspond exactly to the diffraction pattern of the discontinuous helix. The diffraction pattern is formed by the intersection of Q-space with the surface of the Ewald sphere, as shown in Fig. VIII.5, and is also somewhat distorted when it is recorded, as described in Section VIII.3. In particular it will be necessary to tilt the helix, from the perpendicular to the X-ray beam, to measure the intensity at any point on the ζ-axis - other than on the equator. If p and q are very large (compared with the wavelength of the radiation) the intensity distribution of Fig. IX.6 will appear at very low Q values. Then the Ewald sphere is effectively a plane perpendicular to the X-ray beam and the diffraction pattern will closely resemble this figure.

In principle many orders of Bessel function might be expected on each layer line of Fig. IX.6 but, as can be seen from Fig. IX.3, only the lower orders will give rise to appreciable intensity. The $\ell = 0$ layer line corresponds to an origin defined by the undeflected X-ray beam and contains a zero-order Bessel function term. It also coincides with the $\ell = -4$ layer line from the transform of a continuous helix whose origin is placed $2\pi/q$ up the ζ-axis and the $\ell = 4$ layer line from an origin situated $2\pi/q$ down the ζ-axis - since, in this example, p/q = 4. Therefore this composite equator is formed by Bessel functions of order $0, \pm 4$ as well as some much higher, and therefore negligible, orders arising from setting the origin of the continuous helix down at further points, $2\pi/q$ apart, along the ζ-axis. These other orders are only of very low intensity - furthermore their maxima also

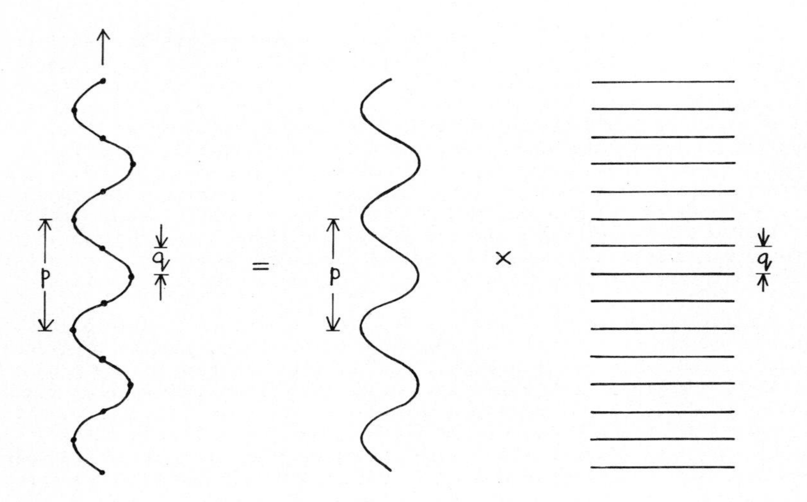

Fig. IX.5. Generation of a discontinuous helix by multiplication.

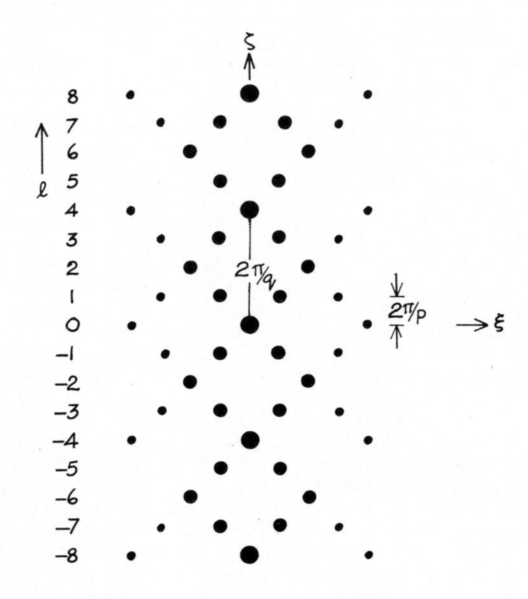

Fig. IX.6. $I(\xi, \ell)$ for a four-fold helix.

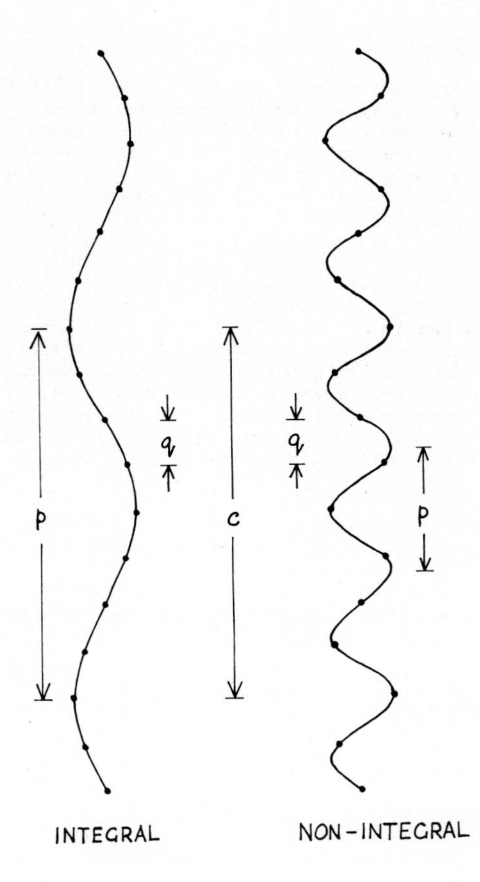

Fig. IX.7. Comparison of integral and non-integral helices.

appear at very high ξ values. Inspection of the figure yields the result that the Bessel function orders, which might appear on the ℓ th layer line, are given by

$$n = \ell + 4m \qquad m = 0, \pm 1, \pm 2 - - - -$$

This selection rule can be generalised to

$$n = \ell + (p/q)\, m$$

provided p/q is an integer.

We now consider the general case where p is not necessarily an integral multiple of q. Figure IX.7 compares two helices, one integral and the other non-integral, with the same true repeat of c. For the integral helix c is identical to the pitch, p, but for the non-integral helix c is greater than p. Now c must be an integral multiple of q and some other integral multiple of p - otherwise the structure could not repeat itself exactly along this distance. Thus

$$c = Nq = tp \qquad \text{(IX.9)}$$

where N and t are integers; for an integral helix t = 1. Both the integral and non-integral helices of Fig. IX.7 have N = 8. According to Section VIII.3 the diffraction pattern consists of layer lines spaced $2\pi/c$ apart - it just happens that for an integral helix c and p are identical. The Fourier transform of the discontinuous helix can still be generated by setting down the transform of a continuous helix at points spaced $2\pi/q$ apart along the ζ-axis - for exactly the same reasons as before. Working out the Bessel function orders that will be expected on each layer line is now more cumbersome because p is not an integral multiple of q. The result is given by the integral values of n obtained from

$$n = (\ell - Nm)/t \qquad m = 0, \pm 1, \pm 2 - - -$$

for an integral helix t = 1 and this result reduces to the selection rule obtained previously.

The general expression for the Fourier transform of a discontinuous helix can now be given. From equation IX.7 the result is

$$\left.\begin{aligned} F(\xi,\psi,\ell) &= \sum_n J_n(\xi r)\, \exp i\, n(\psi + \pi/2) \\ n &= (\ell - Nm)/t \\ m &= 0, \pm 1, \pm 2 - - - \end{aligned}\right\} \qquad \text{(IX.10)}$$

where all m values which do not lead to insignificantly small Bessel function contributions are to be included. This result is simply the sum of all the contributing continuous helix transforms whose origins are spaced $2\pi/q$ apart along the ζ-axis. Figure IX.8 represents the intensity distributions expected from the helices of Fig. IX.7; it shows sections, containing the ζ-axis, of $I(\xi,\ell)$ and is analogous to Fig. IX.6. Bessel function orders contributing to the layer lines, in the non-integral case, are listed in Table IX.1.

TABLE IX.1. Bessel function orders contributing to layer lines for the non-integral helix of Fig. IX.7.

(N=8, t=3)

ℓ \ m	2	1	0	- 1	- 2
8	-	0	-	-	8
7	- 3	-	-	5	-
6	-	-	2	-	-
5	-	- 1	-	-	7
4	- 4	-	-	4	-
3	-	-	1	-	-
2	-	- 2	-	-	6
1	- 5	-	-	3	-
0	-	-	0	-	-
- 1	-	- 3	-	-	5
- 2	- 6	-	-	2	-
- 3	-	-	- 1	-	-
- 4	-	- 4	-	-	4
- 5	- 7	-	-	1	-
- 6	-	-	- 2	-	-
- 7	-	- 5	-	-	3
- 8	- 8	-	-	0	-

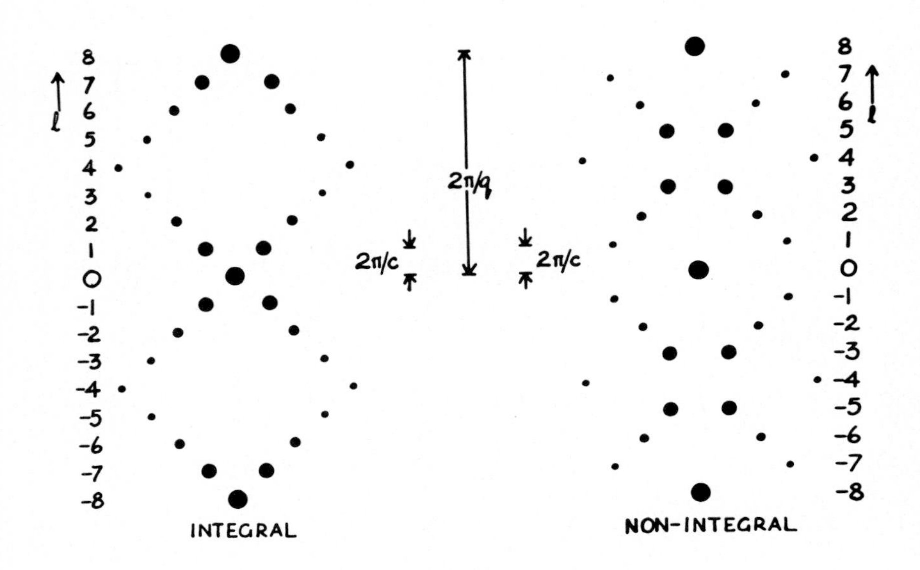

Fig. IX.8. $I(\xi,\ell)$ for integral and non-integral helices.

Values of c, the structural repeat, and q, the axial separation of points, can immediately be deduced from Fig. IX.8. Layer lines are separated, in Q-space, by $2\pi/c$ and hence measurement of this spacing yields c. When measurements are made from a real diffraction pattern, the procedures described in Section VIII.3 have to be adopted. According to Section IX.2, a continuous helix only yields non-zero intensity on the ζ-axis at the origin of Q-space; therefore, a discontinuous helix only yields non-zero intensity along this axis at intervals of $2\pi/q$ in Q-space. If the intensity distribution along the ζ-axis can be recorded, usually from a series of diffraction patterns from tilted specimens, and the distances between non-zero points converted into Q-space, q can be calculated.

Both intensity distributions in Fig. IX.8 have certain similarities; both correspond to the same values of c and q. They can only be distinguished when the orders of the Bessel functions which contribute to each layer line are disentangled. A good idea of the nature of the scattering helix can be gained by inspection of the patterns. To understand the patterns properly r must be known, since the argument of the Bessel functions is ξr, ie. a model for the helix is required. When a model is available, which has the correct values of c and q, the expected diffraction pattern can be calculated and compared with that observed. The problem in interpreting the patterns of Fig. IX.8 is to decide whether c is identical to p (an integral helix) or is greater than p (a non-integral helix). So far we have ignored the hand of the helix. Either helix in Fig. IX.7 could be right- or left-handed, but according to Section III.8, we are unable to determine this hand from the diffraction pattern alone - in the absence of anomalous scattering.

IX.4. Molecular helix

The simplest kind of molecular helix to imagine consists of globular macromolecules arranged along a helical path. Examples are provided by the "thin filaments" of muscle and by the structures of certain viruses. These structures are the same as the discontinuous helix of Fig. IX.5 - except that the points in the figure now represent the positions of the globular sub-units. A molecular helix of this kind may be represented by the convolution of a sub-unit with the discontinuous helix of points.

Consequently the Fourier transform of such a molecular helix is obtained by multiplying the transform of a globular sub-unit, $F(\underline{Q})$, by the transform of the discontinuous helix - given by equation IX.10. $F(\underline{Q})$ can be calculated by the method of Section V.3 or, at low resolution, Section V.4. The Fourier transform of the molecular helix is then given by

$$F(\xi,\psi,\ell) = F(\underline{Q}) \sum_n J_n(\xi r) \exp i\, n(\psi + \pi/2) \qquad \text{(IX.11)}$$

from equation IX.10 which also gives the selection rule for n. If r in equation IX.11 corresponds to the radius of the helix which passes through the centres of the sub-units, then a sub-unit centre must be used as the origin for the calculation of $F(\underline{Q})$. From equations II.5 and IX.11 the intensity distribution is given by

$$\left.\begin{aligned} I(\xi,\ell) &= S(\xi,\ell) F(\underline{Q}) F^*(\underline{Q}) \\ S(\xi,\ell) &= \sum_n J_n(\xi r) \exp\{i\, n(\psi + \pi/2)\} \\ &\quad \times \sum_n J_n(\xi r) \exp\{-i\, n(\psi + \pi/2)\} \end{aligned}\right\} \qquad \text{(IX.12)}$$

Thus arranging the sub-units on a helix introduces an interference function, $S(\xi,\ell)$, which gives the positions in Q-space where intensity is allowed. These allowed positions enable c and q to be calculated as described in Section IX.3. Since the intensity distribution depends on $F(\underline{Q})$, the structure of a sub-unit can be determined, at least in principle, by the methods of Chapter III.

Polymer molecules often form helices. Figure IX.9 shows the structural formula of poly-L-alanine - a single residue of this polymer is boxed in by dotted lines. The molecule can exist in a conformation known as the "α-helix" - the cylindrical polar coordinates of the atoms in a single residue are listed in Table IX.2. If the j th atom in this residue has cylindrical polar coordinates (r_j, θ_j, z_j), the coordinates of the corresponding atom in successive residues are

$$\begin{array}{lll} r_j, & \theta_j + 2\pi t/N, & z_j + q \\ r_j, & \theta_j + 2(2\pi t/N), & z_j + 2q \\ r_j, & \theta_j + 3(2\pi t/N), & z_j + 3q \\ \vdots & \vdots & \vdots \end{array}$$

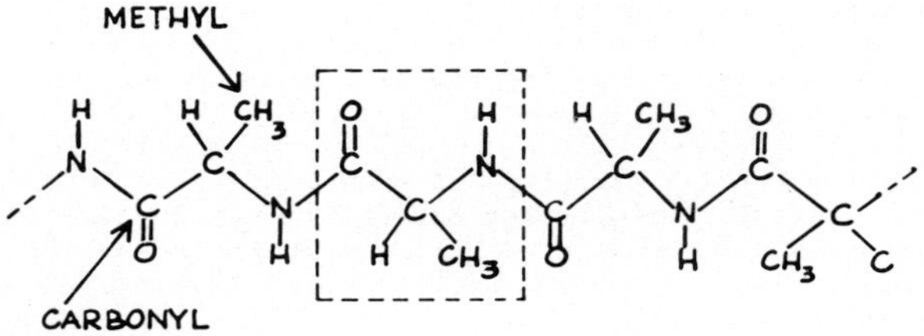

Fig. IX.9. Poly-L-alanine.

TABLE IX.2. Cylindrical polar coordinates of the atoms in a single residue of the α-helix of poly-L-alanine

(Data published by S. Arnott and S.D. Dover, J. Molec. Biol. 30, 209-212, 1967)

	r(nm)	θ(degrees)	z(nm)
C	0.2288	0	0
H (on C)	0.3013	13.54	- 0.0485
C (methyl)	0.3294	- 17.63	0.0808
O	0.1906	78.33	- 0.0761
C (carbonyl)	0.1664	72.41	0.0441
N	0.1548	27.35	0.0906
H (on N)	0.1539	18.57	0.1878

The result for the azimuthal (angular) coordinate follows because the helix turns through $2\pi t$ radians in a repeat distance, c, which contains N residues; the result for the axial coordinate follows directly from the definition of q - the axial rise from one point on a helix to the next equivalent point ie. in the case of a helical polymer it is the axial rise per residue.

Thus the α-helix can be considered as a set of coaxial atomic helices - with one atomic helix for each atom in a residue. The radii of the atomic helices, as well as the relative azimuthal and axial positions of the atoms on them, are given by Table IX.2. All the atomic helices must have the same values of c and q if the relationship between corresponding atoms is to be the same in all residues ie. if covalent bond lengths etc. are to be preserved from one residue to the next. For the α-helix $c = 2.7$ nm and $q = 0.15$ nm. As we might expect, from Section IX.3, these parameters can be measured from a diffraction pattern - further details are given in Section IX.5. Pauling and Corey proposed a model for the α-helix, based on stereochemical principles, in which $p/q = 18/5$; values of N and t then follow from equation IX.9. The coordinates of Table IX.2 were obtained by refining the Pauling and Corey model against the intensity distribution of an X-ray diffraction pattern.

By analogy with equation IX.11 the Fourier transform of a single atomic helix would be

$$F(\xi,\psi,\ell) = f(Q) \sum_n J_n(\xi r) \exp i\, n(\psi + \pi/2) \qquad \text{(IX.13)}$$

where f(Q) is the atomic scattering factor of Section V.2, which is independent of the direction of $\underline{Q}$. But a problem arises when we come to derive an expression for the entire molecular helix - $F(\xi,\psi,\ell)$ is complex and so the correct phase relationships have to be maintained when adding the contributions from the atomic helices ie. each helix must be referred to a common origin. Once the origin is defined for one atomic helix, the same one must be used for all the others. In Table IX.2 the position of a carbon atom was used to define an origin. The j th atom in the residue is related to the azimuthal and axial coordinates of this origin by a rotation, θ_j, and an axial translation, z_j. As in Section V.3, the effect of a translation, when adding atomic scattering factors to obtain a molecular transform, is to multiply $f_j(Q)$ by a term

$$\exp (i\, \underline{r}_j \cdot \underline{Q})$$

where $\underline{r}_j$ is the vector position of the j th atom ie. $\underline{r}$ has the same meaning as in Fig. IX.2. According to equation VIII.10, this term has the form

$$\exp (2\pi\, i\, \ell\, Z_j) = \exp (2\pi\, i\, \ell\, z_j/c)$$

for an axial translation in a one-dimensional crystal; we now have to get the origin of the atomic helix to the same azimuthal coordinate as the origin ie. it has to be rotated by $-\theta_j$ about the z-axis. According to Section II.4, the effect on the Fourier transform is to apply an equal rotation about its ζ-axis ie. ψ in equation IX.13 has to be replaced by $(\psi - \theta_j)$. When equation IX.13 is suitably modified and the contributions of all the atomic helices are added, the Fourier transform of a helical molecule becomes

$$F(\xi,\psi,\ell) = \sum_j \sum_n f_j\, J_n(\xi r_j) \exp i\{n(\psi - \theta_j + \pi/2) + 2\pi\, \ell\, z_j/c\} \qquad \text{(IX.14)}$$

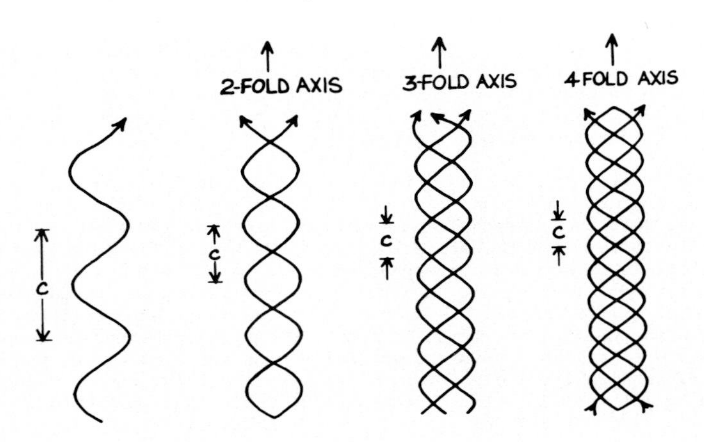

Fig. IX.10. Effect of a u-fold rotation axis of symmetry on the repeat distance of coaxial helices.

Here the summations are over all atoms in a residue and over all the Bessel function orders which make a significant contribution on the layer line - the possible orders being given by the selection rule of equation IX.10.

In some polymers two or more helical chains are wound about a common axis - the much publicised "double helix" of DNA provides an example. An assembly of u coaxial helices is sometimes called a "u-start" helix. Notice that a polymer chain like poly-L-alanine, in Fig. IX.9 is not the same at both ends and so, like an arrow, it defines a direction. Thus a 2-start helix could consist of "parallel" chains, where both point in the same direction, or so-called "anti-parallel" chains, like DNA, which point in different directions. If the constituent chains of a u-start helix are parallel (as opposed to anti-parallel), they may be related by a u-fold rotation axis of symmetry ie. a point on one helical chain may be related to an equivalent point on another chain by a rotation of $2\pi/u$ radians about the helix axis. When c denotes the repeat distance of one of the constituent helices, in isolation, Fig. IX.10 shows that the repeat distance for the entire u-start parallel helix is reduced to c/u - if the chains are related by a rotation axis.

Symmetry can then cause layer lines to be absent from the diffraction pattern. The absence of intensity at otherwise allowed positions in Q-space, as a result of symmetry in real space, is a common feature in X-ray diffraction by crystals - and one-dimensional crystals are no exception. In particular if the parallel chains of a u-start helix are related by a u-fold rotation axis of symmetry, the layer lines expected from a single helix of the same pitch will disappear except when ℓ is a multiple of u. The reason is that the repeat distance is reduced from c to c/u. Consequently the layer line spacing of the diffraction pattern is increased to $2\pi u/c$ in Q-space. The possibility of coaxial, parallel helices, which are related by symmetry, leading to absent layer lines has to be considered when formulating molecular models to explain diffraction patterns.

IX.5. Nematic organisation

The helical polymer molecules considered in the previous section have an overall rod-like shape; as a result they can often be oriented so that their long axes are parallel as shown in Fig. IX.11. Orientation can be achieved by a variety of

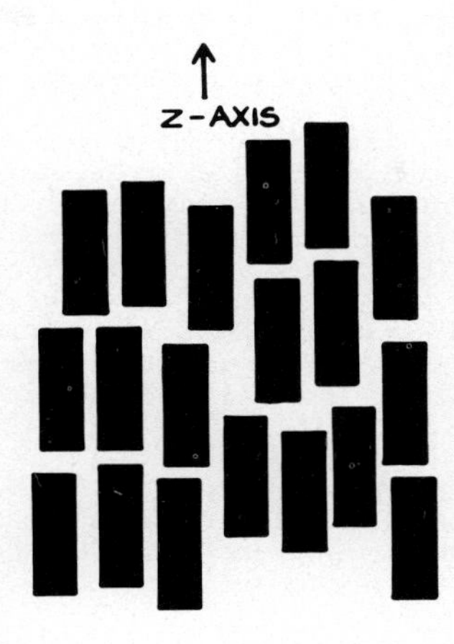

Fig. IX.11. Nematic organisation.

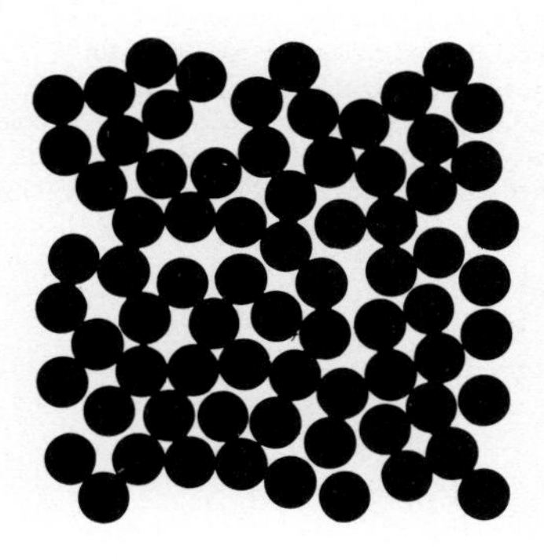

Fig. IX.12. Nematic organisation of rods projected on to a plane perpendicular to the z-axis.

means, such as shearing or extruding concentrated solutions or gels, and is often a desirable feature in synthetic materials because of its effect on mechanical properties. Such biological systems as muscle and tendon also contain oriented polymers for much the same reasons. For studies of molecular structure an oriented sample is preferred to a disoriented assembly of rods because the X-ray diffraction pattern will no longer be spherically averaged.

In many systems the long axes of the rod-shaped molecules are reasonably parallel but the molecules have random rotations about these axes. Furthermore there need be no regular side-to-side spacings between molecules ie. although they are oriented, their arrangement is not crystalline. Then the structure projected on to a plane perpendicular to the z-axis of Fig. IX.11 resembles a random array of discs - as shown in Fig. IX.12. Of course the molecules are not really solid rods, they consist of arrangements of atoms, and so Fig. IX.12 represents only a low resolution view of the projected structure. Because the molecules are not simply rods, the term "parallel" is often used in a specialised and restricted sense.

A polymer chain, like poly-L-alanine in Fig. IX.9, has a direction and can point up or down. Then, when the molecular axes are parallel, molecules themselves could either all point in the same direction ("parallel" - as opposed to "anti-parallel") or point randomly up and down.

A nematic liquid crystal consists of rod-shaped molecules which are oriented, pointing randomly up and down, with no regular side-to-side arrangement ie. in projection it resembles Fig. IX.12. We can therefore call such an arrangement of rods "nematic organisation". The terms "oriented gel" and "uniaxial orientation" are sometimes used for this same arrangement but these terms have become imprecise because both are used in a variety of slightly different ways. In a nematic liquid crystal the rods need not be polymers - they can be relatively small rod-shaped molecules. By analogy with Fig. IX.2, the direction of orientation will be referred to as the z-axis and the ζ-axis is parallel to it in Q-space.

An X-ray diffraction pattern from a nematic provides information about a cylindrically averaged molecule. The reason is that all possible rotations about the long axes of the molecules will be represented in the sample and each of these molecules will scatter X-rays. Suppose, for the sake of argument, that interference between X-rays scattered by different molecules is negligible. Then the system resembles the ideal gas of Section VI.2 except that the intensity is now not spherically averaged but, according to the rotation property of the Fourier transform in Section II.4, is only averaged about the ζ-axis. If $F(\underline{Q})$ is the Fourier transform of a molecule, and there are N molecules in the sample, the intensity distribution is given by

$$I(\underline{Q}) = N < F(\underline{Q})F^*(\underline{Q}) > \qquad \text{(IX.15)}$$

by analogy with the derivation of equation VI.2. Here < brackets > denote a cylindrical average about the ζ-axis, which is equivalent to an average over all ψ values - see Fig. IX.2. Thus, if $\Phi(\psi)$ is some general function of ψ:

$$< \Phi(\psi) > = (1/2\pi)\int_o^{2\pi} \Phi(\psi)\, d\psi$$

The meaning of equation IX.15 is perhaps clearer if it is rewritten, in cylindrical polar coordinates, as

$$I(\xi,\zeta) = N < F(\xi,\psi,\zeta)F^*(\xi,\psi,\zeta) > \qquad \text{(IX.16)}$$

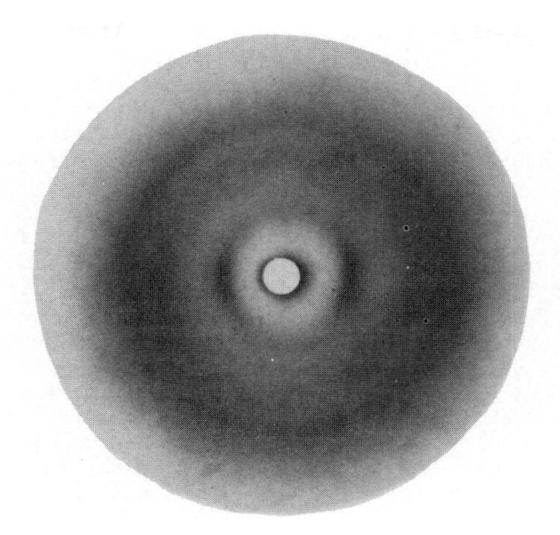

Fig. IX.13. X-Ray diffraction pattern from a lesion in aging human costal cartilage.

If the organisation of rods is truly nematic, ie. there is no regularity in the arrangement of molecules but only orientation, interference effects can be allowed for in much the same way as for a liquid or an amorphous solid. The only difference between equation VI.4 (ideal gas) and equation VII.6 (liquid or amorphous solid) is the appearance of an interference function. By analogy equation IX.16 can be modified to

$$I(\xi,\zeta) = NS(\xi,\zeta) < F(\xi,\psi,\zeta)F^*(\xi,\psi,\zeta) > \qquad \text{(IX.17)}$$

to allow for intermolecular interference. Here $S(\xi,\zeta)$ must be independent of ψ if the intensity distribution is to retain cylindrical symmetry. This independence arises because the arrangement of oriented rods in the nematic is independent of the azimuthal coordinate (θ in Fig. IX.2) of real space.

The diffraction pattern will be formed by the intersection of the cylindrically symmetric intensity distribution, $I(\xi,\zeta)$, with the surface of the Ewald sphere as shown in Fig. VIII.5; note that the X-ray beam is considered to be roughly perpendicular to the z-axis. Because the diffraction geometry is so similar, nematics are considered in the same chapter as one-dimensional crystals. The intensity distribution along the ζ-axis of Q-space provides information about the structure projected on to the parallel z-axis of real space - for exactly the same reason given in Section VIII.4. As before the meridian of the diffraction pattern is still defined to be parallel to the z-axis ie. the direction in which the rods are oriented. The equator is defined to be perpendicular to the meridian and provides information on the structure projected on to a plane perpendicular to the z-axis - once again for exactly the same reason as in Section VIII.4.

The diffraction pattern from an oriented assembly of rod-shaped molecules can be used to determine the direction of orientation. A disoriented array of rod-like molecules gives a diffraction pattern with circular symmetry - like Fig. VII.3 where the macromolecular constituents of the specimen have no preferred orientation. Figure IX.13 does not have circular symmetry because the polymer molecules in this specimen are partially oriented. The direction of preferred orientation is parallel to the meridian of the pattern. Is the meridian vertical or horizontal in Fig. IX.13? Fortunately X-ray diffraction patterns of the polymer in question (collagen) have been obtained from fibres where the orientation was known and we can identify the meridian as vertical. As the molecular structure is also known we could have calculated $I(\xi,\zeta)$ from its coordinates and identified the meridian by comparison of observed and calculated patterns.

It is relatively straightforward to calculate the interference function of equation IX.17 on the equator. We have seen that the equator provides information about the kind of projected structure shown in Fig. IX.12. This structure is like a liquid or an amorphous solid except that it is two-, rather than three-dimensional. It can, therefore, be described by a two-dimensional radial distribution function, $g(r)$, which is independent of direction - except in so far as it is confined to the plane of the figure. By analogy with the three-dimensional definition in Section VII.2, $g(r)$ is simply the probability of finding a centre of one projected molecule at a given distance from another. In the three-dimensional case of a liquid or an amorphous solid, the interference function, $S(Q)$, was related to $g(r)$ by equation VII.6. This equation contains the Fourier transform of

$$\{g(r) - 1\}$$

for a system with spherical symmetry. According to equation II.9, the analogous equation for a system with cylindrical symmetry is

$$S(\xi) = 1 + n_o \int_o^{\infty} 2\pi r \{g(r) - 1\} J_o(\xi r) \, dr \qquad \text{(IX.18)}$$

Here $S(\xi)$ is the equatorial interference function and, since ζ is zero on the equator, it is a function of ξ only; n_o is now the number of discs per unit area. Thus, if a model can be formulated for $g(r)$, the interference function and intensity distribution can be calculated on the equator; the analysis of the equatorial intensity distribution is analogous to the liquid and amorphous solid case of Section VII.4.

There is a simple method of calculating the equatorial Fourier transform at low ξ values. According to Section V.4, the transform is then only sensitive to the projected structure at low resolution and the discs of Fig. IX.12 become a good approximation to the projected structures of the molecules. The Fourier transform of a disc, of radius a, can be calculated from equation II.9, in the same way as the Fourier transform of a sphere was calculated from equation II.8 in Section V.4. In the case of the disc the problem is two-dimensional instead of three-dimensional; the result is

$$F(\xi) = \int_o^a 2\pi r^2 J_o(\xi r) \, dr$$

$$= 2\pi a^2 \{J_1(\xi a)/(\xi a)\} \qquad \text{(IX.19)}$$

In Section VI.3 it was shown that considering roughly spherical molecules to be spheres was an excellent approximation, at low Q values, for calculating the scattered intensity from a system where the arrangement of molecules had spherical symmetry. The approximation given here turns out to be equally good for cylindrically averaged intensity at sufficiently low values of ξ.

Paradoxically, helical molecules, which are themselves one-dimensional crystals, can exhibit nematic organisation ie. their arrangement in space need not be crystalline. Now $F(\xi,\psi,\zeta)$ in equation IX.17 has to be replaced by $F(\xi,\psi,\ell)$ of equation IX.14 because the Fourier transform of a molecule is confined to planes in Q-space which are perpendicular to the ζ-axis. The diffraction pattern is exactly what one would expect of a cylindrically averaged helical molecule modified by an interference function. Note that

$$< F(\xi,\psi,\ell)F^*(\xi,\psi,\ell) > = \sum_n \sum_j J_n^2(\xi r_j)$$

$$+ 2\sum_n \sum_{k>j} J_n(\xi r_k)J_n(\xi r_j) \cos\{n(\theta_k - \theta_j)$$

$$+ (2\pi \ell/c)(z_k - z_j)\}$$

which is, of course, independent of ψ, but does depend on θ_j. The dependence on θ_j arises because intramolecular interference effects depend on the relative azimuthal positions of different atoms in the same molecule. The layer line spacing can be used to measure c, the axial repeat distance of the helices remembering that complications can occur for parallel, coaxial helices - see Section IX.4. If the axial rise per residue, q, is to be measured, the specimen will have to be tilted so that a point on the ζ-axis can touch the surface of the Ewald sphere, as explained in Section VIII.3, and appear on the meridian of the X-ray diffraction pattern. Some idea of the intensity distribution along the ζ-axis can usually be obtained because the intensity distribution of Q-space is blurred out by mosaic spread, as described in Section VIII.8, sufficiently for maxima on this axis to touch the surface of the Ewald sphere. Mosaic spread also leads the intensity which is calculated at a point in Q-space to appear dissipated over an arc on the diffraction pattern.

In principle chiral molecules cannot exhibit nematic organisation; in practice this restriction can be ignored for the very long rods of polymers. Chiral molecules distort the nematic arrangement so that the molecular axes are no longer all parallel; each successive molecular axis acquires a slight rotation or "twist" about the x-axis direction - as shown in Fig. IX.14. The result is called a "twisted nematic" or "cholesteric" liquid crystal; the latter name arose because this phase is commonly formed by esters of cholesterol, although not by cholesterol itself. Helices are chiral and so a helical molecule cannot form a perfectly oriented nematic. However, for a long rod, the rotation between a molecule and its neighbour is extremely small; the net twist of a molecule at the surface of a macroscopic specimen is only a few degrees with respect to a molecule at the centre. Then the effect on the diffraction pattern will be indistinguishable from the mosaic spread of Section VIII.8.

IX.6. Smectic organisation

A smectic liquid crystal has a more ordered arrangement of molecules than a nematic; the least ordered smectics are true one-dimensional crystals. Figure IX.15 shows the arrangement of molecules in a type A smectic; the molecules are oriented and confined to layers. The layers all have the same thickness, c, so that the structure repeats itself regularly in the c-axis direction of the figure. However there is no regularity in the side-to-side arrangement of the rods. Figure IX.12 therefore represents the structure of a type A smectic, projected on to a plane perpendicular to the c-axis, as well as the projected structure of a nematic.

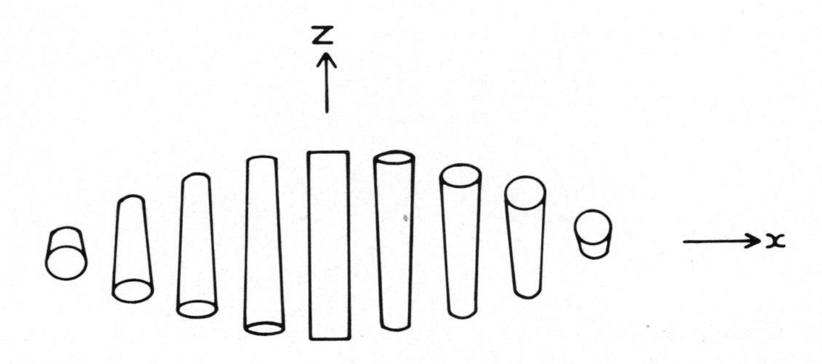

Fig. IX.14. Schematic diagram of the "twist" in a cholesteric liquid crystal.

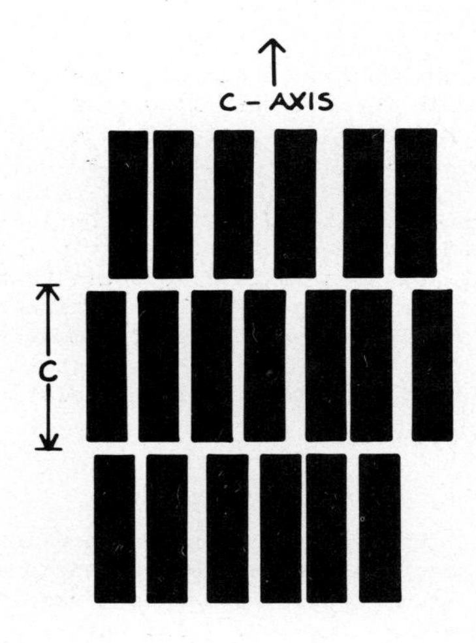

Fig. IX.15. Smectic type A organisation.

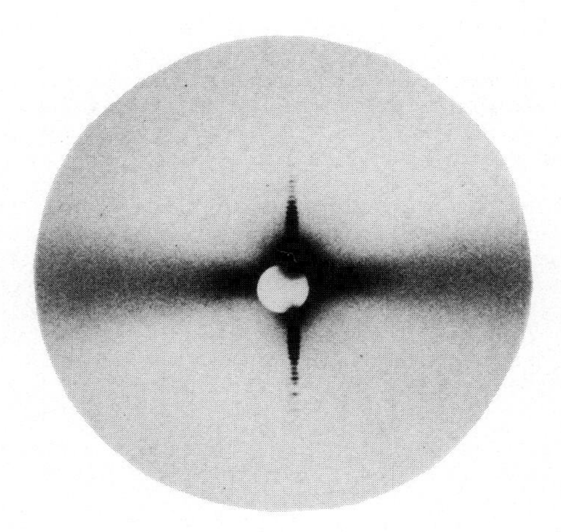

Fig. IX.16. X-Ray diffraction pattern from a fish fin-ray.

In contrast a type B smectic has a regular side-to-side arrangement of molecules within the layers - we are not concerned with such regular structures in this chapter.

Figure IX.16 shows the X-ray diffraction pattern from a system which resembles Fig. IX.15 in having a layered structure but no regular side-to-side arrangement of molecules. The meridian is confined to layer lines which are spaced $2\pi/c$ apart in Q-space. There was no need to tilt the specimen to obtain this pattern because the layer thickness is so great (67 nm) that the pattern appears at very low Q values. At these low Q values the Ewald sphere is essentially a plane section of Q-space, perpendicular to the incident X-ray beam, and so the intensity distribution along the ζ-axis can be recorded without tilting. Notice that the intensity fades at higher ζ values because the planar Ewald sphere approximation breaks down. Then the ζ-axis, along which the intensity happens to be much stronger than elsewhere in Q-space, becomes further from the surface of the Ewald sphere, as shown in Fig. VIII.5; Section VIII.9 gives a detailed explanation of this effect. Also notice that the meridional intensity distribution is not quite symmetrical about the equator, presumably because the specimen was not quite perpendicular to the X-ray beam; this effect is discussed in Section VIII.3. The equator is entirely diffuse because, as in a nematic, it conveys information about an array of projected molecules like Fig. IX.12. Interference effects on the equator can be explained in the same way as for a nematic. Thus Fig. IX.16 corresponds to the diffraction pattern of a true one-dimensional crystal.

IX.7. Summary

Some biological structures consist of globular macromolecules arranged on a helical path; many polymer molecules are helical. If their structures repeat, with a periodicity of c, along the helix axis, their diffraction patterns are confined to layer lines spaced $2\pi/c$ apart in Q-space. Note that c may be greater than the pitch of the helix. A rotation axis of symmetry coincident with the axis of coaxial helices will lead to the absence of some layer lines which would be expected in the diffraction pattern of a single helix.

The diffraction pattern of a helical molecule can also be used to measure q - the axial rise of each "residue" which repeats itself along the helical path. Intensity only appears on the ζ-axis of Q-space, which is parallel to the helix axis, at points spaced $2\pi/q$ apart. The helix axis has generally to be tilted from the perpendicular to the incident X-ray beam for a point on the ζ-axis to appear on a diffraction pattern.

Rod-shaped molecules, eg. helical polymers, can often be induced to assemble with their axes oriented parallel. In the least ordered, oriented array there is no side-to-side regularity in the arrangement of molecules. And they can have random rotations about their long axes. The diffraction pattern then provides information about the structure of a cylindrically averaged molecule; interference effects on the equator can be explained by a two-dimensional version of the theory given in Chapter VII. Such patterns can be used to determine the direction of preferred orientation of polymer chains in materials.

CHAPTER X
Three-dimensional Crystals

X.1. Introduction

A true crystal has a structure which is repetitive in three dimensions. Extension of the theory of diffraction by one-dimensional crystals into three dimensions is all that is required to understand the diffraction patterns from crystals - in principle. In practice X-ray crystallography involves specialised techniques and has some of its own jargon. This specialisation has arisen because X-ray diffraction data can be used to determine the repeating arrangement of atoms in a crystal, and hence molecular structures, with considerable accuracy. Consequently X-ray crystallography has attracted far more attention than any other application of X-ray diffraction.

The principal aim of this chapter is to indicate how the theory which has been developed so far can be applied to analysing X-ray diffraction data from three-dimensional crystals. It is only intended to give a sketchy account of X-ray crystallography as it is actually practised - there are already so many books on this subject that it would be pointless to give yet another account of it. Therefore this chapter serves to link the theory presented in the rest of this book to the many accounts of X-ray crystallography which are already available.

X.2. Intensity distribution

The arrangement of atoms which repeats itself regularly within a crystal is called the "unit cell". Figure X.1 then shows how a crystal can be generated by the convolution of the unit cell contents with a three-dimensional lattice of points. The figure actually demonstrates the properties of a two-dimensional crystal - the behaviour in a third dimension is implied by that in the other two. This simple approach to explaining the properties of three-dimensional crystals will be used throughout this section. Figure X.1 is analogous to Fig. VIII.2 which showed how a one-dimensional crystal could be generated by convolution. The only difference is that, here, the structure is regularly repetitive in three dimensions and, therefore, the lattice consists of an array of points in three dimensions instead of a row of points along a line.

A three-dimensional lattice is defined by three vectors, $\underline{a}$, $\underline{b}$ and $\underline{c}$, as shown in Fig. X.2. In the one-dimensional case of Section VIII.2, the lattice was defined solely by the vector $\underline{c}$ - its modulus, c, was the repeat distance and its direction defined the orientation of the lattice in real space. Here both the

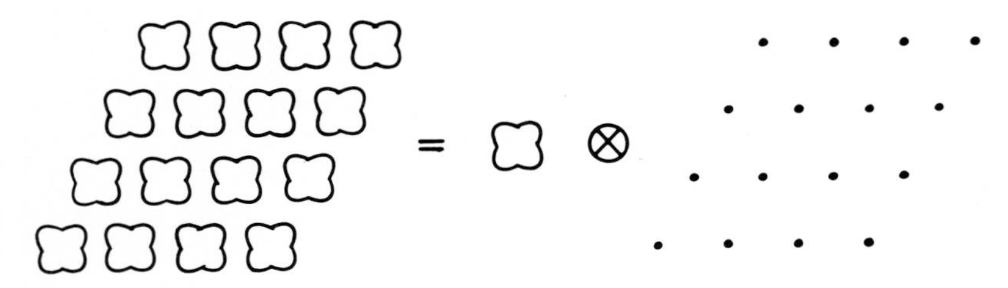

Fig. X.1. Generation of a crystal structure by convolution.

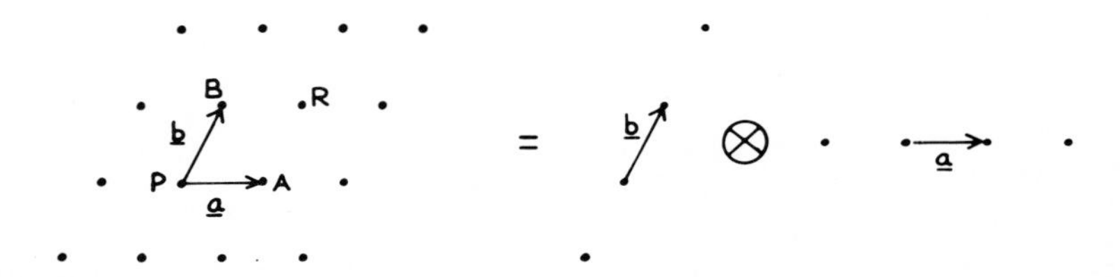

Fig. X.2. Generation of a two-dimensional lattice by convolution.

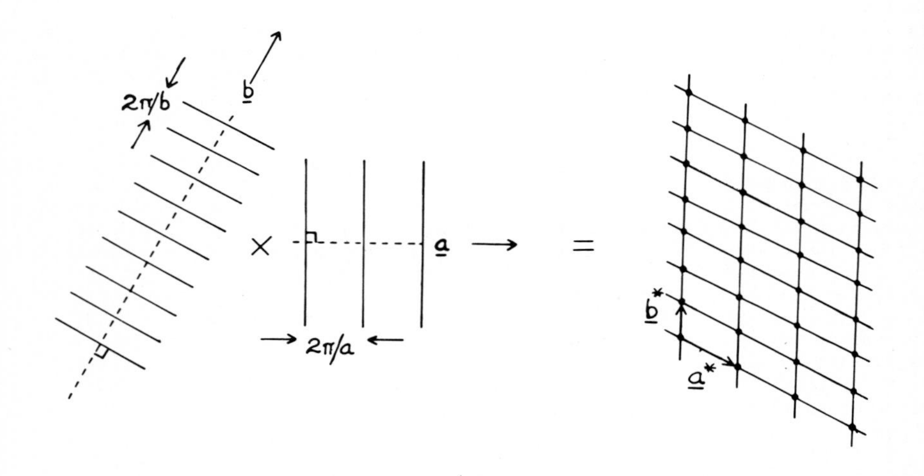

Fig. X.3. Generation of the reciprocal lattice by multiplication.

moduli of $\underline{a}$, $\underline{b}$ and $\underline{c}$, and the angles between them are required to define the lattice. The angles between the pairs of vectors $\underline{a}$ with $\underline{b}$, $\underline{b}$ with $\underline{c}$, and $\underline{a}$ with $\underline{c}$ are often denoted by γ, α and β; note that the usage of α and β is completely different to that in Chapters VIII and IX. In Fig. X.2 the intersection of the two vectors $\underline{a}$ with $\underline{b}$ is taken to be the origin of the unit cell PABR; in three dimensions a vector $\underline{c}$ intersects the other two, also at P, and points out of the plane of this figure - but not necessarily perpendicular to it. The position of the j th atom in the unit cell is given by

$$\underline{r}_j = X_j\underline{a} + Y_j\underline{b} + Z_j\underline{c} \tag{X.1}$$

Here X_j, Y_j and Z_j are translations along the vectors $\underline{a}$, $\underline{b}$ and $\underline{c}$ expressed as fractions of their length; Z_j had exactly the same meaning in Chapter VIII. These fractional translations are often called "fractional unit cell coordinates". It will be convenient to define the vector

$$\underline{X}_j = \begin{bmatrix} X_j \\ Y_j \\ Z_j \end{bmatrix} \tag{X.2}$$

Figure X.2 also shows that a three-dimensional lattice is the convolution of three one-dimensional lattices; Fig. X.3 shows that its Fourier transform is non-zero only at a set of points called the "reciprocal lattice". The real-space lattice in Fig. X.2 can be generated by convolution. Consider a single row of points where each is a distance b apart; the lattice is generated when each of these points acts as an origin for another row of points which are spaced a apart. According to Section II.4, the Fourier transform of the three-dimensional lattice is then given by the product of the transforms of each of the three one-dimensional lattices. Figure X.3 shows that the transforms of the one-dimensional lattices are non-zero only on a set of planes - spaced $2\pi/a$ apart and perpendicular to $\underline{a}$, $2\pi/b$ apart and perpendicular to $\underline{b}$, and $2\pi/c$ apart and perpendicular to $\underline{c}$. This result was derived in Section VIII.3. The product of the three transforms can only be non-zero in those regions of Q-space where each of the three is itself non-zero ie. where their planes intersect. Figure X.3 shows that the planes intersect at a set of points.

The reciprocal lattice can be defined by three vectors, which are conventionally denoted by $\underline{a}^*$, $\underline{b}^*$ and $\underline{c}^*$ as shown in Fig. X.3; in contrast to the usage in the rest of this book, the superscript * has nothing whatever to do with complex conjugates. Angles between the pairs of vectors $\underline{a}^*$ with $\underline{b}^*$, $\underline{b}^*$ with $\underline{c}^*$, and $\underline{a}^*$ with $\underline{c}^*$ are often represented by γ^*, α^* and β^* respectively. In Fig. X.3 $\underline{a}^*.\underline{a}$ is the projection of $\underline{a}^*$ on to the horizontal direction, which equals $2\pi/a$, multiplied by a - the result is 2π. Similarly

$$\left.\begin{array}{ll} \underline{a}^*.\underline{a} = 2\pi & \underline{a}^*.\underline{b} = \underline{a}^*.\underline{c} = 0 \\ \underline{b}^*.\underline{b} = 2\pi & \underline{b}^*.\underline{a} = \underline{b}^*.\underline{c} = 0 \\ \underline{c}^*.\underline{c} = 2\pi & \underline{c}^*.\underline{a} = \underline{c}^*.\underline{b} = 0 \end{array}\right\} \tag{X.3}$$

The dot products whose values are zero arise eg. because $\underline{a}^*$ lies along one of the planes which is perpendicular to $\underline{b}$. Equation X.3 tells us that, from the construction of Fig. X.3, $\underline{a}^*$ is always perpendicular to both $\underline{b}$ and $\underline{c}$, $\underline{b}^*$ is always perpendicular to both $\underline{a}$ and $\underline{c}$, while $\underline{c}^*$ is always perpendicular to both $\underline{a}$ and $\underline{b}$.

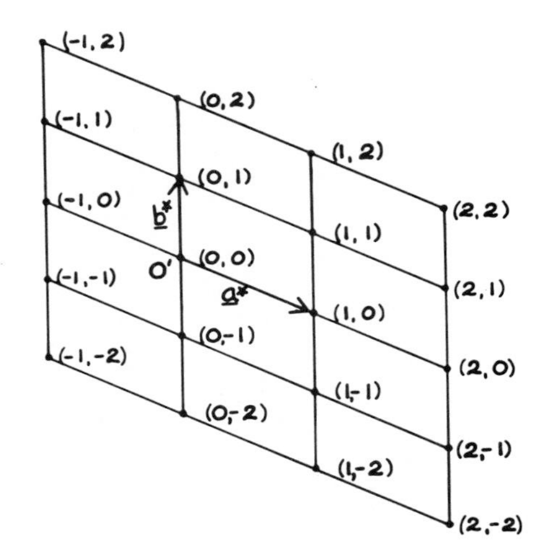

Fig. X.4. Assigning indices to reciprocal lattice points.

Thus these equations may help the reader to picture the direction of $\underline{c}^*$ when Fig. X.3 is extended into three dimensions.

Reciprocal lattice points are specified by the values of three integers - h, k and ℓ. The vectors $\underline{Q}$ which terminate at reciprocal lattice points are given by

$$\underline{Q} = h\underline{a}^* + k\underline{b}^* + \ell\underline{c}^* \tag{X.4}$$

where h, k and ℓ are integral and all three have a value of zero at the origin of Q-space, O'. Figure X.4 shows a two-dimensional example where only two of these integers, h and k, are required. It will often be convenient to define the vector

$$\underline{h} = \begin{bmatrix} h \\ k \\ \ell \end{bmatrix} \tag{X.5}$$

whose components must be integral if the Fourier transform of the lattice is to be non-zero. Note that

$$\begin{aligned} Q &= \left| h\underline{a}^* + k\underline{b}^* + \ell\underline{c}^* \right| \\ &= \{(h\underline{a}^* + k\underline{b}^* + \ell\underline{c}^*).(h\underline{a}^* + k\underline{b}^* + \ell\underline{c}^*)\}^{\frac{1}{2}} \\ &= (h^2a^{*2} + k^2b^{*2} + \ell^2c^{*2} + 2hka^*b^* \cos\gamma^* \\ &\qquad + 2k\ell b^*c^* \cos\alpha^* + 2h\ell a^*c^* \cos\beta^*)^{\frac{1}{2}} \end{aligned} \tag{X.6}$$

Equation X.6 is the general expression which gives the values of Q corresponding to the reciprocal lattice points; it therefore applies to so-called "triclinic" crystals which are characterised by

$a \neq b \neq c$

$\alpha \neq \beta \neq \gamma \neq 90^\circ$

Most crystals have geometrically more regular unit cells and equation X.6 can then be simplified.

What is the value of the Fourier transform of the lattice at the reciprocal lattice points? According to Fig. VIII.3 the peaks in the Fourier transform of a one-dimensional lattice, of effectively infinite extent, all have the same height. Consequently the value of the Fourier transform of a three-dimensional lattice will be the same at all the reciprocal lattice points. This value will be very large, so that the crystal acts as an amplifier for the Fourier transform of the unit cell contents, at the reciprocal lattice points, as discussed in Section VIII.4. Its absolute value will be of no concern since we shall not be concerned with absolute measurements of intensity in this book. Consequently the Fourier transform of the lattice may be assigned a value of unity, on some arbitrary scale, at each of the reciprocal lattice points.

The Fourier transform of the crystal can be obtained by multiplying the transform of the electron density distribution in a unit cell by the transform of the lattice. This result arises because the crystal structure can be generated by the convolution of the unit cell contents with the lattice. The Fourier transform of the electron density in a unit cell is given by equation V.4. Although this equation has been taken to represent the Fourier transform of a molecule, it is valid for any assembly of atoms provided their relative positions are fixed - as is clear from the derivation of Section V.3. Since the transform of the lattice has a value of unity at the reciprocal lattice points and zero elsewhere, we simply have to calculate the values of the "molecular transform" of equation V.4 at the reciprocal lattice points in order to obtain the Fourier transform of a crystal. From equations X.1 and X.4:

$$\begin{aligned} \underline{r}_j \cdot \underline{Q} &= (X_j\underline{a} + Y_j\underline{b} + Z_j\underline{c}) \cdot (h\underline{a}^* + k\underline{b}^* + \ell\underline{c}^*) \\ &= 2\pi(hX_j + kY_j + \ell Z_j) \\ &= 2\pi\underline{h} \cdot \underline{X}_j \end{aligned} \qquad \text{(X.7)}$$

Here the second line follows from equation X.3 and the third line from equations X.2 and X.5. Then, from equations V.4 and X.7, the Fourier transform of the crystal is given by

$$F(\underline{h}) = \sum_j f_j \exp 2\pi i \underline{h} \cdot \underline{X}_j \qquad \text{(X.8)}$$

where the summation has to be taken over all the atoms in a unit cell. Now the Fourier transform is only non-zero at those points in Q-space which are given by integral components of $\underline{h}$ - it can, therefore, be written as $F(\underline{h})$ instead of as $F(\underline{Q})$. Remember that the atomic scattering factor, f_j, depends on Q which, in general, is related to the components of $\underline{h}$ by equation X.6.

The values of the Fourier transform of a crystal, $F(\underline{h})$, at the reciprocal lattice points are often called "structure factors" by crystallographers - although sometimes this term is applied to the modulus of $F(\underline{h})$. Scattered intensity is confined to reciprocal lattice points and, according to equation II.5, is given by

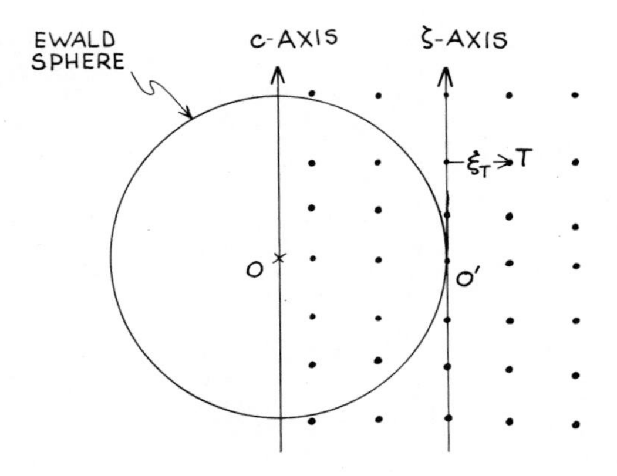

Fig. X.5. Section through the Ewald sphere showing reciprocal lattice points.

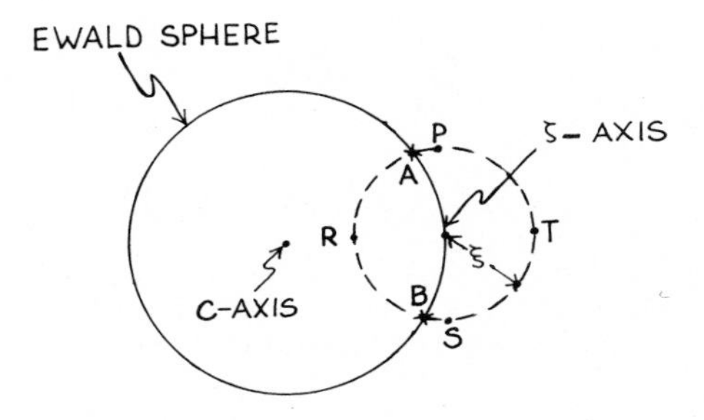

Fig. X.6. Section through points in Q-space with the same ζ values.

$$I(\underline{h}) = F(\underline{h})F^*(\underline{h}) \qquad (X.9)$$

where $F(\underline{h})$ is given by equation X.8 and $F^*(\underline{h})$ is its complex conjugate. The point on a diffraction pattern where a value of $I(\underline{h})$ is recorded is often called a "reflection". This nomenclature arises from Bragg's law, of Section VIII.6, which is represented by equation VIII.6. For a three-dimensional crystal an equation like equation VIII.6 can be derived by considering X-rays to be reflected by planes of atoms in the crystal - hence the term "reflection".

X.3. Diffraction geometry

If the crystal remains stationary, Fig. X.5 shows that few reciprocal lattice points will touch the surface of the Ewald sphere. An exception occurs if the unit cell is very large - then the reciprocal lattice points will be so closely spaced that some are bound to touch its surface. But the diffraction pattern will depend on the precise position of the crystal and may be difficult to interpret. Figure X.5 is a section through the Ewald sphere when $\underline{c}$ is perpendicular to the incident X-ray beam. The cylindrical polar coordinates (ξ,ψ,ζ) of a point in Q-space are defined in Fig. IX.2. In the rest of this section, and indeed in the remainder of the book, the ζ-axis will be defined to be parallel to $\underline{c}$. Figure X.6 shows a section of the Ewald sphere which is perpendicular to the ζ-axis, ie. to the plane of Fig. X.5, and so passes through points in Q-space with the same values of ζ. If few reciprocal lattice points touch the surface of the Ewald sphere, there will be zero intensity on nearly all of the diffraction pattern.

An informative pattern can be recorded if the crystal in Figs. X.5 and X.6 is rotated about the direction of $\underline{c}$ - which is generally called its "c-axis". According to Section II.4 $F(\underline{h})$ and, therefore, $I(\underline{h})$ are rotated about the parallel ζ-axis of Q-space. Reciprocal lattice points then pass through the surface of the Ewald sphere and the corresponding values of $I(\underline{h})$ appear as spots on the diffraction pattern. Consider the reciprocal lattice points, like P, R, S and T of Fig. X.6, which have the same values of ξ and ζ. They will all pass through the same points, A and B, on the surface of the Ewald sphere. Consequently the dependence of the scattered intensity on ψ cannot be detected. A diffraction pattern recorded in this way is called a "rotation pattern" of the crystal.

Rotation patterns can be easily understood if they are converted into Q-space coordinates; the result is called a "reciprocal lattice rotation diagram" and an example is given here for an orthorhombic unit cell. An orthorhombic cell has the following properties:

$$a \neq b \neq c$$

$$\alpha = \beta = \gamma = 90^\circ$$

This example is a compromise - it is not so complicated as the general, or triclinic, case but it is not too regular in its dimensions. As before the crystal is considered to rotate about its c-axis so that its Fourier transform is rotated about the parallel ζ-axis. For an orthorhombic cell $\underline{c}^*$ is parallel to $\underline{c}$ ie. it is coincident with the ζ-axis. The c-axis can be located in practice by adjusting the tilt of the crystal until the four quadrants of its rotation pattern are identical - more details later. Note that, according to Section VIII.3, the intensity distribution along the ζ-axis will not be recorded when this axis is perpendicular to the incident beam; the position defined by the undeflected beam is an exception but its intensity will be too great to record with the rest of the pattern - according to Section III.7.

If the diffraction pattern is recorded on a flat film, the method described in Section VIII.3 can be used to convert the coordinates of each point on the film into coordinates in Q-space. However, rotation patterns are usually recorded on a cylindrical film which is coaxial with the rotation axis. The transformation between film and Q-space coordinates, and the corresponding Bernal chart, are then given in Vol.II of the "International Tables for X-Ray Crystallography" (see BIBLIOGRAPHY Section 2). But "International Tables" uses different units for Q-space (corresponding to the Ω-space of Section IV.3) - the conversion factor can be found by comparing its equations for a flat film with those in Table VIII.1. Notice also that in "International Tables" the ζ-axis is defined more generally than here; in the tables it is parallel to the axis about which the crystal is rotated. In this book the crystal is considered to be rotated about its c-axis but this is not necessarily the case - it could equally well be rotated about the a-axis or the b-axis.

The positions of the spots on the orthorhombic rotation diagram will now be derived. Since $\underline{a}$, $\underline{b}$ and $\underline{c}$ are mutually perpendicular, in this case, so are $\underline{a}^*$, $\underline{b}^*$ and $\underline{c}^*$. Also $\underline{a}^*$ is now parallel to $\underline{a}$, $\underline{b}^*$ is parallel to $\underline{b}$, and $\underline{c}^*$ is parallel to $\underline{c}$ so that equation X.3 gives

$$\left.\begin{aligned} a^* &= 2\pi/a \\ b^* &= 2\pi/b \\ c^* &= 2\pi/c \end{aligned}\right\} \qquad \text{(X.10)}$$

From equation X.10

$$\zeta = \ell c^* \qquad \text{(X.11)}$$

since the ζ-axis now lies along the direction of $\underline{c}^*$. Since α, β and γ are all equal to 90°, so are α^*, β^* and γ^*. Therefore equation X.6 becomes

$$Q = (h^2a^{*2} + k^2b^{*2} + \ell^2c^{*2})^{\frac{1}{2}} \qquad \text{(X.12)}$$

and it also follows that

$$\begin{aligned} \xi &= (Q^2 - \zeta^2)^{\frac{1}{2}} \\ &= (h^2a^{*2} + k^2b^{*2})^{\frac{1}{2}} \end{aligned} \qquad \text{(X.13)}$$

where the final line of equation X.13 follows from equations X.11 and X.12. If a, b and c are known, equations X.10, X.11 and X.13 can be used to calculate where the reciprocal lattice points will appear - information about ψ is lost when the crystal is rotated because all reciprocal lattice points with the same values of ξ and ζ pass through the same two points on the surface of the Ewald sphere.

Figure X.7 shows the orthorhombic, reciprocal lattice rotation diagram. In this example a* has been chosen to be greater than b* ie. a is less than b. Note that the positions of the spots on the rotation diagram are the same in all four quadrants. The same will be true of the diffraction pattern since the recording process distorts all four quadrants equally. This diagram conforms to the general rule, discussed earlier in this section, that a reciprocal lattice point appears in two places on the diffraction pattern. Previously this result was obtained pictorially - now we can see that the two positions correspond to the positive and negative square roots in equation X.13. Reciprocal lattice points denoted by (h,k,ℓ), $(-h,k,\ell)$, $(h,-k,\ell)$ and $(-h,-k,\ell)$ all pass through the points marked with

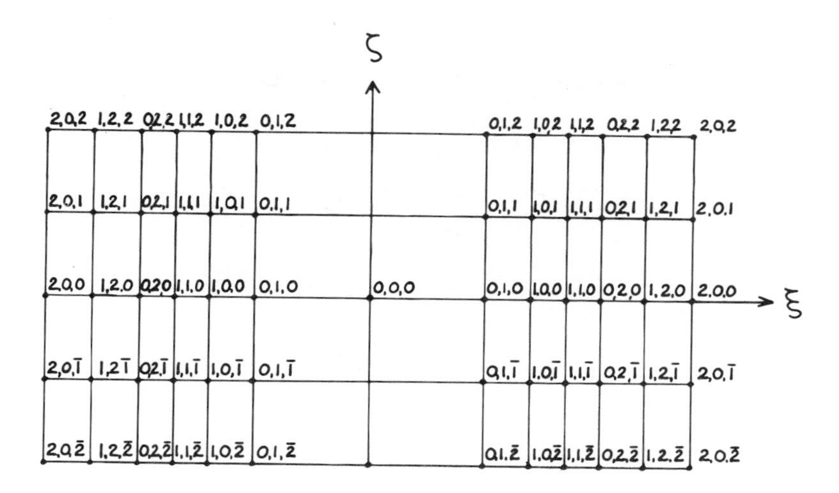

Fig. X.7. Positions of spots (h,k,ℓ) on an orthorhombic reciprocal lattice rotation diagram (negative values of ℓ are denoted by $\bar{\ell}$)

values of (h,k,ℓ) in Fig. X.7 - because all four sets of indices lead to the same values of ξ and ζ according to equations X.11 and X.13.

In practice the positions of the spots on an X-ray diffraction pattern are used to measure the unit cell dimensions; then Fig. X.7 is obtained experimentally and the construction of the reciprocal lattice is reversed to obtain the required results. The first step is to assign values of h, k and ℓ to each spot ie. to "index" the pattern. Indexing usually proceeds by trial-and-error. From Fig. X.7 it might appear straightforward but several snags can occur; two examples will be given. For a tetragonal crystal, which is the same as the orthorhombic case except that a and b are equal, pairs of spots like (1,2,1) and (2,1,1) will overlap. If γ^* is not 90°, ξ is a function of γ^* - which complicates the positions of the spots. These results can be derived from equation X.6, once the reciprocal lattice has been constructed, but they are summarised by equations in Vol.II of "International Tables". When the many spots have been indexed, and their Q values measured, there are more than enough data to determine the six unknowns - a^*, b^*, c^*, α^*, β^* and γ^*. In our orthorhombic example equations X.11 and X.13 would be used to obtain a^*, b^* and c^* from ξ and ζ. The construction of the reciprocal lattice can then be reversed - if necessary with the aid of equations in Vol.II of "International Tables". In our example a, b and c can be calculated from a^*, b^* and c^* using equation X.10. If equations X.10, X.11 and X.13, which represent a special case, are to be used it must first be recognised that the unit cell is indeed orthorhombic ie. that α, β and γ are all equal to 90°.

What can be predicted about the intensities of the spots? In our orthorhombic example of Fig. X.7 the intensity of each spot marked (1,2,1) is really the sum of the intensities at the (1,2,1), (-1,2,1), (1,-2,1) and (-1,-2,1) reciprocal lattice points. The intensities of the same reciprocal lattice points contribute to both the (1,2,1) spots so they must have the same net intensity. Similarly the spots marked (1,2,-1) are really a composite of the intensities at the

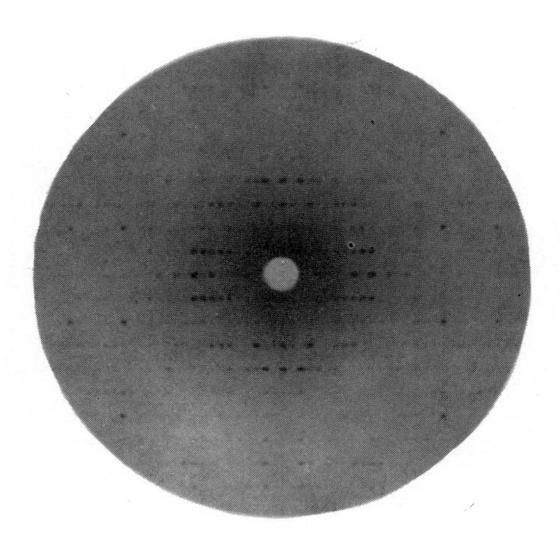

Fig. X.8. X-Ray diffraction pattern of a crystal of the enzyme 6-phosphogluconate dehydrogenase, extracted from sheep liver, obtained by the precession method (taken by Dr. M.J. Adams).

(-1,-2,-1), (1,-2,-1), (-1,2,-1) and (1,2,-1) reciprocal lattice points. Thus the spots marked (1,2,1) are the sum of a set of I(h) values and those marked (1,2,-1) are a sum of the corresponding I(-h) values. According to equation X.4, if h corresponds to a vector Q, then -h corresponds to -Q and according to Section III.8 I(Q) and I(-Q) are equal. Consequently the spots marked (1,2,1) and (1,2,-1) have the same intensity. When this argument is applied to other pairs of spots, (h,k,ℓ) with $(h,k,-\ell)$, it can be seen that the intensity distribution of all four quadrants of the rotation pattern is the same. This result is true whatever the shape of the unit cell. If the intensity were not the same in all four quadrants, ie. if I(h) were not everywhere equal to I(-h), anomalous scattering, discussed in Section V.5 would be appreciable - then, according to Section V.6, the chirality of the unit cell contents could be determined.

Overlap of spots on the diffraction pattern can be prevented by oscillating the crystal back and forth through a small angle. The resulting "oscillation patterns" allow the intensities of each of the spots which were superimposed in the diffraction pattern to be measured separately. If the crystal were oscillated through $\pi/4$ radians from its position in Fig. X.6, only the reciprocal lattice point marked P would contribute to A and only that marked S would contribute to B. A second oscillation pattern, after the crystal had been rotated by $\pi/2$ radians, would allow points like R and T to pass through the surface of the Ewald sphere.

There are many other methods for recording X-ray diffraction patterns from crystals. The precession method is particularly interesting because it records undistorted sections through Q-space. It achieves this result by precessing the film over the surface of a sphere, which is concentric with the Ewald sphere, while the crystal rotates. By collecting a series of sections, the intensity at each reciprocal lattice point can be recorded. An example of a precession photograph is shown in Fig. X.8. For accurate intensity measurements counters are often used instead of photographic film to record the diffraction pattern. But for crystals with large unit cells the spots are too closely spaced for a counter to measure the intensity of each separately - so film then has to be used.

X.4. Determination of crystal structures

We have seen that the positions of the spots which constitute the diffraction pattern of a crystal can be used to measure its unit cell dimensions; their relative intensities can be used to measure the electron density distribution within a unit cell. Since the Fourier transform of a crystal is non-zero only at a set of points, the integral of equation II.7 can be written as a summation. The result is

$$\rho(\underline{r}) = \sum_{\underline{h}} F(\underline{h}) \exp(-i\, \underline{r}.\underline{Q}) \tag{X.14}$$

where the integral components of $\underline{h}$ range in value from minus to plus infinity ie. information from an infinite number of diffraction spots is required. From equations X.7 and X.14

$$\rho(\underline{X}) = \sum_{\underline{h}} F(\underline{h}) \exp(-2\pi i\, \underline{h}.\underline{X}) \tag{X.15}$$

Note that the electron density distribution need only be calculated within a single unit cell. However, the positions of the atoms are unknown - the aim of the experiment is to find them. Therefore $\rho(\underline{X})$ has to be calculated at a large number of closely spaced points in the hope that atomic positions can be inferred from regions of high electron density.

The application of equation X.15 raises two problems: the phase problem of Section III.2 and the unavailability of an infinite number of spots. In essence the phase problem arises because, in general, $F(\underline{h})$ is a complex number and $I(\underline{h})$, which is measured experimentally, can provide only its modulus. Methods for solving the phase problem were given in Sections III.4 and III.5. Note that the simple "direct method" provided as an example in Section III.4 is not applicable to a crystal - because its Fourier transform is not a smoothly varying function but is restricted to points in Q-space. Direct methods for crystals are more subtle and are described in the book by Woolfson (see BIBLIOGRAPHY Section 12). What are the consequences of the number of spots being limited? According to Section III.6, if only a limited extent of Q-space, ie. a limited number of spots, can be recorded, the result is that the resolution of the technique is limited. However, the theoretical resolution is perfectly adequate to determine atomic positions with the X-ray wavelengths generally used in diffraction experiments. Protein crystals are usually imperfect, in the sense described in the one-dimensional case of Section VIII.7. The intensity of their diffraction spots then decreases with increasing Q and it may not be feasible, or even possible, to record the intensities of those which convey high resolution information. Stereochemical information, ie. covalent bond lengths and angles, van der Waals' radii etc., then has to be combined with the calculated electron density distribution in order to find the positions of atoms. Remember that, if the resolution is severely limited, spurious peaks may appear in $\rho(\underline{X})$ as demonstrated in Section III.6.

Why is X-ray crystallography such a useful technique for determining molecular structure? The principal reason is that the diffraction data do not correspond to a molecular structure which has been spherically or cylindrically averaged - in contrast to the systems discussed in Chapters VI to IX. Of course the intensity, $I(\underline{h})$, at each reciprocal lattice point has to be measured separately; the intensities of the composite spots in a rotation pattern provide cylindrically averaged information - because the dependence of the scattered intensity on ψ is lost. X-Ray crystallography has three further advantages. Accurate intensity measurements are possible because the intensity distribution consists of discrete spots on a background which, for a perfect crystal, is featureless. Also, for a perfect crystal, intermolecular interference effects do not lead the intensity to fall off with increasing Q - in contrast to the case of a liquid or amorphous solid. Measurement of high resolution data is then straightforward. Although real crystals are not perfect, they are often sufficiently near perfect for these advantages to apply. The third advantage is that, in practice, it is virtually always feasible to apply the deductive methods of structure determination, described in Section III.4; a crystal structure can, therefore, be determined with a high degree of certainty.

If accurate atomic positions are to be determined, it is necessary to correct the measured intensities of the diffraction spots for three different effects. First the intensity of each spot has to be divided by the polarisation factor, p, defined in equation I.5, for the reason given in Section II.3. Next the Lorentz correction, of Section VIII.9, has to be applied; further details of the Lorentz factor are given in Vol.II of the "International Tables for X-Ray Crystallography". Finally, if very accurate results are required, an absorption correction must be applied. This correction recognises that X-ray beams scattered with different Q values, ie. at different scattering angles, pass through differing thicknesses of the specimen. Then, according to equation V.14, absorption can reduce the intensity of each by a slightly different factor. Note that, further to Section II.3, we are only concerned with correction factors which are a function of Q - since we are not concerned with measurements on an absolute scale (see also Section VI.2). Methods which can be used to place intensity measurements on an absolute scale are described in most books on X-ray crystallography.

Equation X.15 can be easily modified to allow for the kind of crystal imperfection described in Section VIII.7; crystallographers usually attribute this imperfection to thermal vibrations but Section VIII.7 suggests that this is often an oversimplification. Suppose that the atoms in a crystal vibrate independently, then the Fourier transform of each atom, ie. its atomic scattering factor, has to be multiplied by a function with the form of $\tilde{D}(Q)$ in equation VIII.16. The result is that equation X.8 has to be replaced by

$$F(\underline{h}) = \sum_j f_j(Q) \exp(-Q^2\sigma_j/2) \exp 2\pi i\, \underline{h}.\underline{X}_j \qquad (X.16)$$

In reality the atoms do not vibrate independently but neither do they vibrate in unison; the book by James (see BIBLIOGRAPHY Section 2) gives a more detailed account of thermal vibrations in crystals. Furthermore there is no reason to suppose that, at a given temperature, the amplitude of an atomic vibration will be the same in all directions in a crystal. Crystallographers sometimes make an empirical attempt to deal with this latter problem by replacing $Q^2\sigma_j/2$ in equation X.16 with

$$b_{11}h^2 + b_{22}k^2 + b_{33}\ell^2 + b_{12}hk + b_{23}k\ell + b_{31}h\ell$$

where b_{11}, b_{22} - - - are called "anisotropic temperature factors". The form of the dependence on h, k and ℓ arises from using equation X.6 to expand Q.

Refinement of the atomic positions, inferred from the calculated electron density distribution, forms the final stage of a structure determination. The parameters to be varied in this refinement may include the position $\underline{X}_j$ and thermal vibration parameters (corresponding to either σ_j or the anisotropic temperature factors) of each atom, and often an overall scale factor (see Section VI.2). This refinement is best carried out in Q-space - for the reason given in Section III.6. When the number of spots whose intensity has been measured greatly exceeds the number of variable parameters, very precise structural information can be obtained. If the data do not extend to sufficiently high Q values for individual atomic positions to be resolved, this approach is not valid. Atomic positions then have to be varied subject to the condition that they are related by reasonable bond lengths and angles while non-bonded atoms do not approach closer than the sum of their van der Waals' radii; furthermore it would be unreasonable to attempt to obtain individual thermal vibration parameters for each atom, in such cases, so that overall values have to be assigned to entire molecules.

X.5. Symmetry

The unit cell possesses symmetry if the positions of all its atoms can be generated by the operation of symmetry elements on a subset of these positions. This subset, from which all the other positions can be generated, is called the "asymmetric unit" of the unit cell. Symmetry elements reflect, rotate or translate a point to an equivalent point. Some symmetry elements actually perform a combination of two of these operations eg. a right-handed u-fold screw axis translates a point through a distance p/u, where p is a repeat distance in the axial direction, and rotates it by $2\pi/u$ radians. Thus points which are related by a u-fold screw axis lie on a u-fold helix of pitch p.

The properties of symmetry elements will be illustrated by using a 2-fold rotation axis as an example. If the j th atom in a unit cell has fractional unit cell coordinates (X_j, Y_j, Z_j), a 2-fold rotation axis which is coincident with the c-axis generates another, equivalent atom at $(-X_j, -Y_j, Z_j)$. If the position of the original atom is denoted by $\underline{X_j}$, of equation X.2, the operation of the rotation axis can be represented by the matrix

$$\begin{bmatrix} -1 & 0 & 0 \\ 0 & -1 & 0 \\ 0 & 0 & 1 \end{bmatrix}$$

since

$$\begin{bmatrix} -1 & 0 & 0 \\ 0 & -1 & 0 \\ 0 & 0 & 1 \end{bmatrix} \begin{bmatrix} X_j \\ Y_j \\ Z_j \end{bmatrix} = \begin{bmatrix} -X_j \\ -Y_j \\ Z_j \end{bmatrix}$$

Only 230 combinations of symmetry elements are allowed. The reason is that the combination of two or more symmetry elements can generate further symmetry elements and only certain combinations are then compatible. Suppose that a unit cell has

$$\alpha = \beta = \gamma = 90^{\circ}$$

with 2-fold rotation axes along the c- and b-axes; the effect of these is to multiply the vector $\underline{X_j}$, which defines the position of the j th atom, first by

$$\begin{bmatrix} -1 & 0 & 0 \\ 0 & -1 & 0 \\ 0 & 0 & 1 \end{bmatrix}$$

and then by the analogous matrix for a rotation axis coincident with the b-axis - which is

$$\begin{bmatrix} -1 & 0 & 0 \\ 0 & 1 & 0 \\ 0 & 0 & -1 \end{bmatrix}$$

Therefore the net operation is represented by multiplying $\underline{X}_j$ by the product of the two matrices. The result is

$$\begin{bmatrix} -1 & 0 & 0 \\ 0 & -1 & 0 \\ 0 & 0 & 1 \end{bmatrix} \begin{bmatrix} -1 & 0 & 0 \\ 0 & 1 & 0 \\ 0 & 0 & -1 \end{bmatrix} = \begin{bmatrix} 1 & 0 & 0 \\ 0 & -1 & 0 \\ 0 & 0 & -1 \end{bmatrix}$$

which, by analogy with the matrices from which it is generated, represents a 2-fold axis along the a-axis direction; the two 2-fold axes generate a third. Similar arguments show that any one of the three can be generated from the other two. This kind of behaviour is not confined to 2-fold rotation axes but applies to any combination of symmetry elements acting on a body. Any set whose elements can be generated from each other in this way is called a group. Given the kinds of symmetry elements which can be applied to a repetitive structure like a crystal, it turns out that there are 230 possible groups of elements - called "space groups". These 230 space groups include the case with no symmetry elements.

The symmetry exemplified by certain space groups can only be observed if the unit cell of the crystal has a particular shape. In the example of the previous paragraph it was presumed that

$$\alpha = \beta = \gamma = 90^\circ$$

Actually no other combination of values for α, β and γ is compatible with the only symmetry elements of the unit cell being three, non-parallel 2-fold rotation axes. Then the unit cell must have mutually perpendicular edges ie. its shape is orthorhombic ($a \neq b \neq c$), tetragonal ($a = b \neq c$) or cubic ($a = b = c$). If the three rotation axes were neither parallel nor perpendicular, their rotations would be coupled so that they would be incapable of acting according to their definitions ie. they could not exist.

If the atoms in the unit cell are related by certain combinations of symmetry elements some spots, which might be predicted to appear in the diffraction pattern, will actually be absent. Coaxial, parallel helices, in Section IX.4, provide an example. There the presence of a u-fold rotation axis along the helix axis (which we now recognise as a screw axis) led to the absence of layer lines unless ℓ was an integral multiple of u. In general, symmetry may lead spots to be systematically absent unless the three integral components of $\underline{h}$ (ie. h and k as well as ℓ) obey similar rules. The pattern of absences for each of the 230 space groups is listed in Vol.I of the "International Tables for X-Ray Crystallography"; these tables allow the space group symmetry of the unit cell to be determined.

Symmetry can be used to simplify expressions for $F(\underline{h})$ in equations X.8 and X.16 as well as for $\rho(\underline{X})$ in equation X.15. In the case of $F(\underline{h})$ the summation of equations X.8 and X.16 need only extend over the atoms in a single asymmetric unit; the contributions of the other atoms can be generated by symmetry. The usual practice is to modify the form of $F(\underline{h})$ to allow for these contributions. Similarly it is only necessary to calculate $\rho(\underline{X})$ over an asymmetric unit and the electron density in the rest of the unit cell can be generated by symmetry. Often many of the values of $F(\underline{h})$ which contribute to the summation of equation X.15 will be zero. It is known, for a given space group symmetry, which values of the components of $\underline{h}$ will lead to a zero value for $F(\underline{h})$; consequently the expression for $\rho(\underline{X})$ can be modified so that the values of $F(\underline{h})$ which are systematically zero do not have to be included. Special forms of equations X.8 and X.15, for each of the 230 space groups, are listed in Vol.I of "International Tables". Note that equations X.8 and X.15 are actually the forms appropriate for the space group with no symmetry elements.

X.6. Summary

A crystal structure is regularly repetitive in three dimensions; the repeating unit is called its "unit cell". The Fourier transform of a crystal is confined to points in Q-space - the "reciprocal lattice points". Each reciprocal lattice point can be specified by the values of three integers - denoted by h, k and ℓ.

The intensity distribution of the diffraction pattern from a single stationary crystal corresponds to an insufficient extent of Q-space to provide much useful information. To record a useful diffraction pattern the crystal has to be rotated or oscillated - in some techniques the film (or other detector) moves also. The diffraction pattern is confined to discrete spots. Each spot corresponds to one, or more, reciprocal lattice points and can be assigned values of h, k and ℓ by inspection - precise details depend on the experimental technique used to record the pattern. When the spots have been "indexed" in this way, and their positions converted into Q-space, the unit cell dimensions can be calculated.

If the phase problem can be solved, the electron density distribution within a unit cell can be calculated from the intensities of the spots which constitute the diffraction pattern. Positions of atoms within the unit cell, and hence molecular structures, can be inferred from the calculated electron density distribution - the coordinates of the inferred atomic positions have to be refined to obtain accurate molecular models.

CHAPTER XI

Crystalline Powders and Crystalline Fibres

XI.1. Introduction

The aim of this chapter is to describe the X-ray diffraction effects observed from two systems which although crystalline do not consist simply of a single crystal - crystalline powders and crystalline fibres. A particle in a finely ground powder is typically less than 1 μm across; yet in many powders each particle is a crystal. There will be so many of these tiny crystals in a macroscopic specimen that each has an effectively random orientation in space. Crystalline fibres contain oriented polymer molecules but, in contrast to the nematic and type A smectic organisations of Chapter IX, their molecules have a regular side-to-side arrangement.

No new theory is introduced in this chapter - all that is needed is to show how the theory of Chapters VIII to X can be applied to these systems. Before starting Section XI.3 it might be useful to look back at Fig. IX.2. This figure defines the cylindrical polar coordinates (ξ,ψ,ζ) which are used to define the position of a point in Q-space - these coordinates are particularly helpful when describing the diffraction properties of crystalline fibres. As we shall see, the diffraction patterns from crystalline powders and crystalline fibres are of considerable practical value.

XI.2. Powder method

Perhaps the major application of the powder method is the identification of unknown substances. A sample of the crystalline substance which is to be identified is ground into a fine powder so that it consists of a vast number of micro-crystals; these micro-crystals will have random orientations in space.

The diffraction properties of this crystalline powder can be understood using the theory developed for a mosaic crystal in Section VIII.8. In three dimensions the crystallites, of which the mosaic crystal is composed, will be tiny three-dimensional crystals. A perfect three-dimensional crystal consists of these crystallites stacked together neatly, like a pile of bricks. Mosaic spread consists of the crystallites being randomly tilted from perfect alignment about the three axes of a Cartesian coordinate system. A powdered specimen then represents an extreme case of mosaic spread where the tilt about each axis can range from zero to 2π radians; with such a large number of tiny crystals, every possible tilt will be

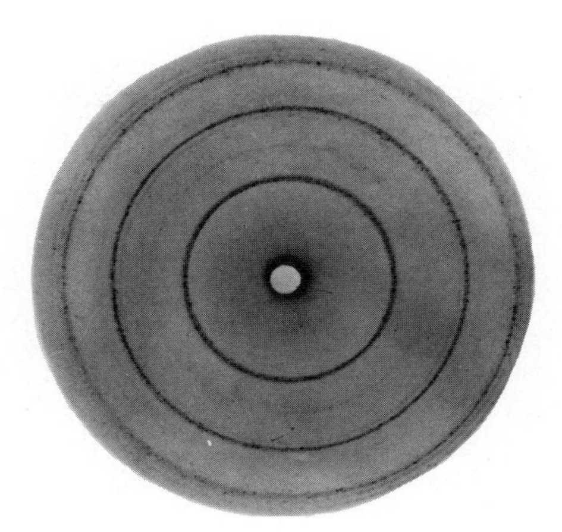

Fig. XI.1. X-Ray diffraction pattern of a micro-crystalline deposit in a bone marrow biopsy; the positions of the rings and their relative intensities showed that the micro-crystals consisted of calcium oxalate.

equally represented.

A diffraction pattern from a crystalline powder consists of sharp, concentric rings - as illustrated by the example of Fig. XI.1. According to Section VIII.8, the intensity at a reciprocal lattice point is distributed over part of the surface of a sphere by mosaic spread. The radius of the sphere is given by the value of Q corresponding to the reciprocal lattice point and the area over which the intensity is spread depends on the degree of tilting of the crystallites in the mosaic crystal. In this extreme case, where the tilt can be as great as 2π radians, and all possible tilts are equally represented, the intensity of a reciprocal lattice point will be evenly distributed over an entire spherical surface in Q-space. The intensity distribution in Q-space is, therefore, confined to a series of concentric spherical shells. Each shell intersects the Ewald sphere around a circle - as described in Section IV.2. Consequently the diffraction pattern consists of concentric rings. The radius of each ring depends on the value of Q for the corresponding reciprocal lattice point and hence, according to equation X.6, on the unit cell dimensions. Relative intensities of the rings depend on the positions of the atoms within a unit cell - according to equations X.8 and X.9.

Thus the radii and relative intensities of the rings can be used as a "finger print" to identify both the chemical constitution and the crystal form of an unknown substance. Several compilations of radii, converted into Q-space, and relative intensities are available from powder patterns of standards which consist of known substances in known crystal forms. A substance is identified when the measurements from its powder pattern coincide with those from a standard.

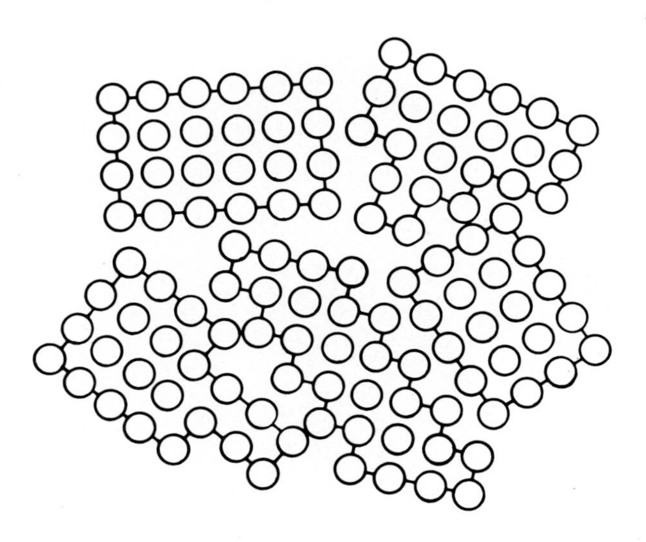

Fig. XI.2. Molecular packing in a crystalline fibre.

XI.3. Crystalline fibres

Polymer molecules have helical symmetry when they are oriented. Consider a single polymer molecule in an oriented assembly. If the molecule is capable of being oriented as described in Section IX.5 it must have a rod-like shape. In such a molecule each successive residue must have the same conformation. A single residue which had a very different conformation would produce a kink - such a kink would prevent the molecules from becoming oriented. Similarly a succession of residues with wildly different conformations would lead to a polymer chain which was in no sense rod-like and would be incapable of orientation. If the residues in a rod-like polymer molecule have the same conformation they will be related by a screw axis as defined in Section X.5 ie. the polymer molecule has helical symmetry. Such molecules need not resemble cork-screws - although some do eg. the α-helix of poly-L-alanine in Section IX.4. All that is implied by the description "helical" is that the cylindrical polar coordinates of corresponding atoms, in successive residues, are related as described in Section IX.4 - a 2-fold helical molecule, ie. one in which residues are related by a 2-fold screw axis, is distinctly ribbon-like in shape.

Many polymers form crystalline fibres, which consist of assemblies of crystallites, when their molecules are oriented. One of the unit cell axes, conventionally designated the c-axis, is defined by the periodicity of the molecular helix. A single polymer chain is so long that it may pass through several crystallites. Since the molecules are oriented, and their helix axes define the c-axis, the c-axes of the crystallites must then be, at least roughly, parallel. However there is no reason for any such continuity between crystallites in other directions. Figure XI.2 represents the molecular packing in a crystalline fibre projected on to a plane perpendicular to the c-axis direction. The projected molecules are represented as discs and the boundaries of crystallites are marked. Here the moduli of $\underline{a}$ and $\underline{b}$ are equal, the two vectors are perpendicular and there is only one molecule in the unit cell - but none of these properties is a necessary feature of crystalline fibres. The important point is that the crystallites are randomly rotated about their c-axes. Since the crystallites are very small, there will be very many in a macroscopic specimen; all rotations, about the c-axis, between zero and 2π radians will then be equally represented.

A crystalline fibre is, therefore, a very special kind of mosaic crystal. According to Section VIII.8, its diffraction pattern is then formed by adding the intensity contribution of each crystallite. Each crystallite rotation about the c-axis is associated with an equal rotation of its Fourier transform about a parallel axis in Q-space - from Section II.4. As in Section X.3 the ζ-axis of Q-space is defined to be parallel to the c-axes of the crystallites. The rotated Fourier transform, multiplied by its complex conjugate, gives the corresponding intensity distribution for the rotated crystallite. When the contributions from all the crystallites are added, the resultant intensity distribution is cylindrically averaged ie. the dependence of the scattered intensity on ψ is lost.

At this point it may help to compare the diffraction properties of crystalline fibres with those of crystalline powders. In a crystalline powder the crystallites have all possible orientations in space. As a result the intensity at a reciprocal lattice point is distributed over the surface of a sphere of radius Q ie. the intensity distribution is spherically averaged. In a crystalline fibre the crystallites have all possible rotations about their c-axes. When the contributions to the scattered intensity from all the crystallites are added, the net intensity is cylindrically averaged ie. if $I(\xi,\psi,\zeta)$ is the intensity distribution scattered by a single crystallite, expressed as a function of cylindrical polar coordinates, the intensity distribution from the entire fibre is given by

$$I(\xi,\zeta) = (1/2\pi)\int_0^{2\pi} I(\xi,\psi,\zeta)\, d\psi$$

(see Section IX.5). Consequently a reciprocal lattice point at (ξ,ψ,ζ) has its intensity distributed around a circle of radius ξ whose centre lies on the ζ-axis, at a distance ζ from O', the origin of Q-space; the surface enclosed by this circle is, of course, perpendicular to the ζ-axis.

Diffraction geometry from a crystalline fibre is just the same as for a rotating single crystal. The reason is that all reciprocal lattice points with the same values of ξ and ζ are distributed around the same circle in Q-space. They will then pass through the same two points on the surface of the Ewald sphere, exactly as in Section X.3, when the diffraction pattern is formed. Alternatively this result could have been explained by appealing to the inability of X-ray diffraction to distinguish between systems which are averaged in space and time - as in Section VII.1. The single crystal rotation method gives a diffraction pattern which is cylindrically averaged in time; the diffraction pattern from a crystalline fibre is cylindrically averaged, about the same axis, in space.

Figure XI.3 provides an example of a diffraction pattern from a crystalline fibre; the spots can be indexed as described in Section X.3. Notice two features of these spots. One: they are rather broad - indicating, from Fig. VIII.3, that there is a limited number of unit cells in a crystallite. Two: they are distinctly arc-shaped - indicating, from Section VIII.8, that the c-axes of the crystallites are only roughly parallel ie. the polymer molecules are imperfectly oriented. Arcs reduce the resolution of diffraction patterns from crystalline fibres. According to Section VIII.8, the arc length increases with Q - then the intensity of a spot is dissipated over an increasingly large area of photographic film ie. the intensity falling on a unit area of film is decreased. Consequently it will become impossible to detect the intensity of spots whose positions correspond to high Q values - according to Section V.4 this limits the resolution of the diffrac-

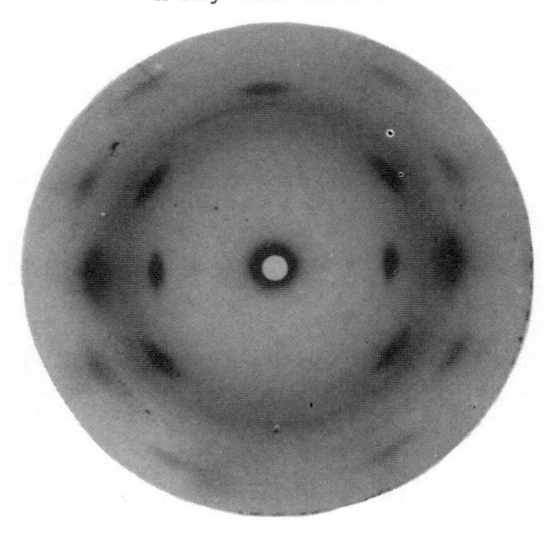

Fig. XI.3. X-Ray diffraction pattern of stretched rubber; the c-axis was tilted slightly from the perpendicular to the incident X-ray beam (taken by K.E. Davies).

Fig. XI.4. A "statistical" crystallite structure.

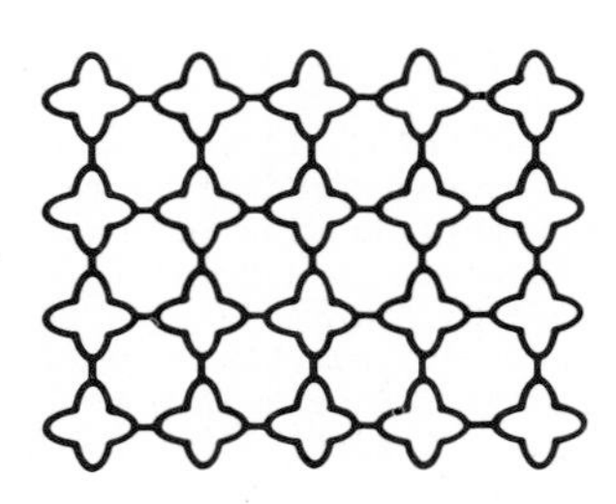

Fig. XI.5. A perfect crystallite structure.

tion pattern.

Interference effects, between X-rays scattered by different molecules are much easier to calculate than was the case for nematic organisation of oriented polymers - see Section IX.5. However, a complication, which frequently arises, is that crystallites have "statistical" structures. Figure XI.4 provides a schematic example - in projection on to a plane perpendicular to the c-axis. Here each molecular site is randomly occupied by a molecule with one of two possible orientations. Another particularly common kind of statistical crystallite structure consists of molecules which point randomly up or down the c-axis direction. Statistical structures also involve molecules with different translations along the c-axis. And each site can be randomly occupied by more than two different possibilities. Each molecule scatters X-rays and the intensity at each reciprocal lattice point refers to the structure of an average unit cell. Furthermore diffuse intensity appears between the reciprocal lattice points - its appearance will now be explained. Figure XI.5 could be considered as a "perfect" crystallite structure. Then the protruberances on some of the projected molecules in Fig. XI.4 are slightly displaced from the ideal positions of Fig. XI.5 - according to Section VIII.7, diffuse intensity then appears between reciprocal lattice points. When indexing diffraction patterns from crystalline fibres it is important not to attribute maxima in the diffuse intensity to intensity at reciprocal lattice points.

XI.4. Interpretation of diffraction patterns from crystalline fibres

The intensities of the spots in the diffraction pattern of a crystalline fibre can be used to determine the three-dimensional structure of its polymer molecules. At first sight there may appear to be three difficulties. One is low resolution - but this can easily be overcome, as described in Section X.4, by supplementing the diffraction data with stereochemical information in the form of covalent bond lengths, bond angles etc. Another is that, in practice, the deductive methods of determining structure, described in Section III.4, are rarely applicable so that the trial-and-error approach of Section III.5 has to be used instead; fortunately, as we shall see in the next paragraph, this approach is particularly effective for helical molecules. The third apparent difficulty is that the intensities of the diffraction spots provide information about a cylindrically averaged structure ie. the dependence of the intensity scattered by a crystallite on ψ is lost. However, if the trial-and-error approach is to be used, cylindrical averaging presents no real problems.

Development of models for the three-dimensional structures of polymer molecules in crystalline fibres is reasonably straightforward - because the molecules are helical. When a helix is packed in a crystallite its Fourier transform is amplified at the reciprocal lattice points but becomes zero elsewhere. Thus its transform is sampled at those points in Q-space which correspond to the reciprocal lattice points; the relative intensities at these points are unchanged by molecular packing - except for interference between X-rays scattered by different helices in the same unit cell. Therefore the intensity distribution along the ζ-axis can be used to determine the axial rise per residue of the helix, as in Section IX.4, and the layer line spacing provides a value for the helical repeat, c. Remember that the intensity distribution along the ζ-axis has to be inspected on diffraction patterns from suitably tilted fibres. But, because the orientation will usually be imperfect, a good idea of this distribution can be gained without tilting - as described in Section VIII.9. Also the relative intensities of the spots provide a valuable clue as to whether integral or non-integral helical models should be considered; Section IX.3 provides further details.

Models have to be produced for the helical molecule and for the packing of helices in a crystallite. Before a molecular model can be developed, the structural formula of the polymer, like that of poly-L-alanine in Fig. IX.9, must be known. The conformation of a residue is then adjusted, incorporating the stereochemical information which is to supplement the diffraction data, so that successive residues join together to form a helix of the required dimensions. This procedure is usually carried out by computer. Next a model is developed for the packing of the helices in a unit cell; the unit cell dimensions can be determined from the positions of the diffraction spots as soon as the pattern is indexed - as in Section X.3. Remember that statistical crystallite models, described in Section X.7, may be appropriate. Next the intensity distribution, calculated from the coordinates of the model, is compared with that observed - structure determination, including refinement, proceeds as in Section III.5. Nowadays refinement is usually carried out in a computer, where the variable parameters of the model are automatically adjusted to obtain the best fit between calculated and observed intensity distributions.

During the trial-and-error structure determination, the hand of the helical molecules is usually determined. Few polymers have achiral residues. A right-handed molecular helix of chiral residues is not the mirror image of a left-handed helix consisting of the same residues. Denoting the chirality of a residue by L, and that of its mirror image by D, the mirror image of a left-handed helix of L residues is a right-handed helix of D residues. But if our polymer is known to contain L residues, there are no valid models containing D residues. Any competing left- and right-handed helical models must both contain L residues and will not be related as object to mirror image. The arguments of Section III.8 are not applicable to these models and the better of the two can be determined as in Section III.5.

An essential feature of the trial-and-error method is the calculation of the expected intensity distribution from the coordinates of a model for the structure - how is this calculation performed? When a molecular model has been built and a scheme for packing the molecules in a unit cell developed, a set of trial coordinates will be available for all the atoms in a unit cell. If these are expressed as fractional unit cell coordinates, equation X.16 can be used to calculate $F(\underline{h})$. This calculation is performed for each reciprocal lattice point which contributes to the diffraction pattern. Equation X.9 is then used to calculate $I(\underline{h})$ from $F(\underline{h})$. Values of $I(\underline{h})$ for overlapping reciprocal lattice points are added; overlapping points have the same values of ξ and ζ. The result is a cylindrically averaged intensity distribution which can be compared with that observed. If a statistical crystallite model is being developed, the summation of equation X.16 will include atomic coordinates for every molecular position which can appear in the structure; the atomic scattering factor, f_j, then has to be multiplied by the probability that an atom will appear at this position in a single unit cell - this probability is sometimes termed the "occupancy" of the site.

Equation X.16 is still valid, even though the molecules are helical. There is no real need to develop expressions which take into account the helical symmetry of the molecules in a unit cell. Such expressions will reduce the number of atoms included in the summation - since all the atomic positions in a unit cell can usually be generated from the coordinates of the atoms in a single residue of the helical molecule. However, using a modern computer, computation and summation of the terms in equation X.16 is extremely rapid - so very little time will be saved if an alternative expression is used.

XI.5. Summary

In the diffraction pattern from a crystalline powder the intensity associated with a reciprocal lattice point is distributed over a circle. The pattern then consists of concentric rings - the radii of these rings converted into Q-space, and their relative intensities, are characteristic of a particular chemical substance in a particular crystal form. Comparison of the radii and intensities, measured from the diffraction pattern of an unknown powder, with the results from powders of known composition, provides a useful method of chemical analysis.

Many oriented polymer samples are crystalline fibres. These fibres consist of crystallites whose c-axes are defined by the axes of the helical polymer chains. Crystallites have random rotations about their c-axes; diffraction geometry for a crystalline fibre is then identical to that for a rotating single crystal. A diffraction pattern from a crystalline fibre can be used to determine the conformation of its polymer chains.

CHAPTER XII

Relationship to Microscopy

XII.1. Introduction

One purpose of this chapter is to show how the Abbe theory of the microscope provides a useful aid to understanding the formation and analysis of X-ray diffraction patterns. The Abbe theory emphasises the relationship between an object, its diffraction pattern and an image of the object - a relationship which is fundamental to X-ray diffraction analysis. Formation of a diffraction pattern is an intermediate stage in the formation of an image by a microscope. Since microscopes can image irregular specimens, the illusion, often fostered by X-ray diffraction texts, that regularity in the object is necessary for the formation of a diffraction pattern, is immediately dispelled. This chapter, therefore, summarises much that has already appeared in earlier chapters by presenting the material in a rather different way; the theory of the microscope provides a useful model for X-ray diffraction analysis.

Another purpose of this chapter is to indicate how the methods used in the analysis of X-ray diffraction patterns can be applied to electron microscopy and the attendant technique of electron diffraction. Details of image analysis and the reconstruction of three-dimensional images from electron micrographs are beyond the scope of this book - as are details of the various diffraction techniques associated with electron microscopy. But much of the underlying theory of these techniques is essentially the same as that used in the analysis of X-ray diffraction patterns. The Abbe theory of the microscope emphasises the relationship of electron microscopy to electron diffraction and X-ray diffraction. Once this relationship is recognised, much of the contents of this book could be considered as an introduction to the theory which is applied to the analysis of electron micrographs.

XII.2. Fourier transformation and microscopy

Figure XII.1 shows how a microscope forms a magnified image, A'B', of an object, AB. A point source of light is positioned at the focus of the condenser lens, in order to provide a parallel beam of light. Since monochromatic X-rays are usually employed in diffraction experiments, the source will be assumed to be monochromatic here - to make comparison easier. The object scatters the incident light beam. All scattered beams which are parallel are focused at the same point by the objective lens; the pattern of scattered light, which is formed at the back

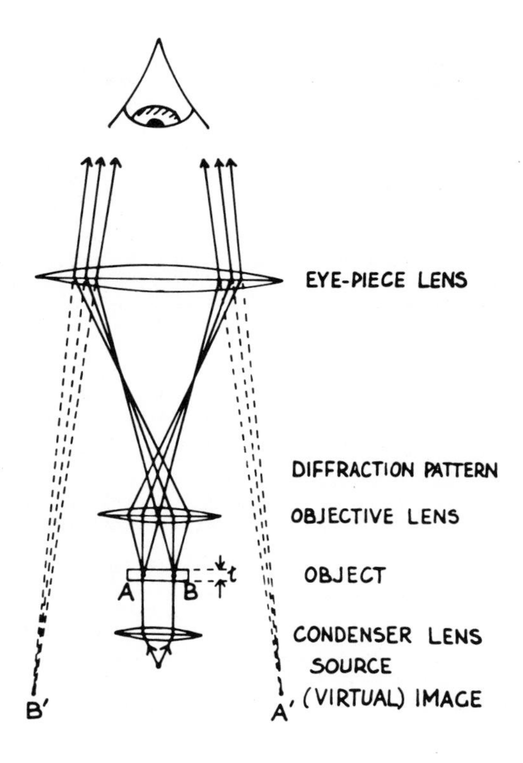

Fig. XII.1. Image formation in the microscope.

focal plane of the objective, is called the Fraunhofer diffraction pattern of the object. Throughout this chapter "Fraunhofer diffraction pattern" will be abbreviated to "diffraction pattern".

A microscope forms an image in two stages: first a diffraction pattern is formed and then an image. Consider the formation of the diffraction pattern first. The amplitudes and phases of the scattered waves depend on the density of scattering matter in the object. When the principles of X-ray scattering were discussed, in Chapters I and II, it emerged that the amplitudes and phases could be calculated from the electron density distribution, $\rho(\underline{r})$; when light scattering is being considered it is usual to think in terms of molecular polarisability - the relationship between molecular polarisability and refractive index is described in Section II.6. In Fig. XII.1 the objective lens takes all waves which are scattered in the same direction and through the same angle, ie. all waves with the same $\underline{Q}$, and focuses them at a point. Each point in the back focal plane of the objective lens then corresponds to a point in Q-space ie. the diffraction pattern is focused in this plane.

Since the diffraction pattern is not recorded in the microscope, the phase relationship between the scattered waves is maintained. There will be a phase associated with the waves focused at each point on the back focal plane of the objective. If a photographic film were used to record the diffraction pattern, all phase information would be lost - as described in Section III.2. But, in the microscope, the waves continue through this plane and the correct phase relationships between waves focused at different points, ie. at different points in Q-space, are maintained.

The microscope then synthesises an image from the scattered light. In Fig. XII.1 the scattered waves pass through the back focal plane of the objective lens and eventually all the waves emanating from A meet at one point and all those from B meet at another. Thus the objective lens acts as a magnifying glass to form first a diffraction pattern and then an image. In the microscope the eye-piece lens magnifies this intermediate image to form the final image A'B'. A lens is needed if all the waves scattered by A are to meet at a single point ie. no image can be formed without a lens.

An image is formed from a diffraction pattern by inverse Fourier transformation. According to Section II.3 the diffraction pattern is represented by the Fourier transform, $F(\underline{Q})$, of $\rho(\underline{r})$. From Section II.4, an image of the object can then be obtained by inverse transformation of $F(\underline{Q})$. Note that here "diffraction pattern" refers to the waves passing through the back focal plane of the objective - information is conveyed both by their amplitudes and phases. Thus the microscope transforms $F(\underline{Q})$ rather than $I(\underline{Q})$. Consequently the transformation yields an image, as in Section II.4, rather than the autocorrelation function, as in Section III.3.

According to Sections V.4 and III.6, the resolution of the microscope depends on the extent of Q-space which is used to synthesise an image. If the objective aperture of the microscope is decreased, waves scattered at high Q values cannot appear in the diffraction pattern and so the resolution of the microscope is diminished; if the aperture is so reduced that the extent of Q-space is severely limited, spurious detail may appear in the image - as described in Section III.6. Section III.6 also shows that the maximum value of Q which is accessible with a given wavelength, λ, is

$$Q_{max} = 4\pi/\lambda \tag{XII.1}$$

Thus if the wavelength of the light is reduced, a greater extent of Q-space can be synthesised into an image and the resolution of the microscope is improved.

Resolution will now be treated more quantitatively. Suppose we wish to image a row of holes, which are equally spaced a distance c apart, in an opaque screen. What wavelength is required for each hole to be separated in the image ie. for the holes to be resolved? If the diffraction pattern were recorded, at the back focal plane of the objective, it would consist of lines spaced $2\pi/c$ apart in Q-space - according to Section VIII.3. Each line is assigned an integer index, ℓ. The line which passes through the centre of the diffraction pattern, perpendicular to the plane of Fig. XII.1, has a zero value of ℓ. If the adjacent lines can also be seen the diffraction pattern appears as a set of lines; the two nearest adjacent lines correspond to ℓ values of plus and minus unity. The extent of Q-space which is required to detect them is

$$2(2\pi/c) = 4\pi/c$$

This same extent of Q-space must be synthesised into an image if the nature of the object is to be revealed - a greater extent provides more detail. Equation XII.2 summarises the result:

$$4\pi/c \leqslant Q_{max} \qquad \text{(XII.2)}$$

From equations XII.1 and XII.2 the condition that the holes are resolved is

$$c \geqslant \lambda$$

ie. the theoretical resolution of the microscope is of the same order as the wavelength of the light.

There are specialised techniques in microscopy for adjusting contrast by manipulating the diffraction pattern before the image is synthesised. In "dark-field" microscopy the contrast is enhanced by preventing waves scattered at very low Q values from contributing to the image. If Fig. III.4 represents the density of scattering matter in an object, Fig. III.5 represents F(Q) and Fig. III.8, which is the inverse transform omitting F(Q) at low Q values, then represents the amplitude of the light across the image formed by the dark-field microscope. Figure III.8 shows that the dark-field technique subtracts a constant background from this amplitude - and hence from the intensity of the image. Suppose that an object consists of bodies which scatter very little more light than their surrounding medium - the contrast in the image will be poor. But the dark-field technique can be used to subtract a constant background from this image which corresponds to the intensity contribution from the medium surrounding the bodies. They will then appear as light areas on a black background. Unfortunately Fig. III.8 also shows that spurious detail, which was not present in the original object, can appear in the dark-field image.

What is the effect of the thickness, t, of the object in Fig. XII.1? Suppose that the microscope is required to image details whose dimensions are of the order of λ ie. it is operating near the limit of its resolution. In Section IV.2 it was shown that the region of Q-space which can be explored by scattered radiation is restricted to the surface of the Ewald sphere. Thus only that part of Q-space which lies on this spherical surface is used to synthesise an image of the object. If a plane section of Q-space, which contained its origin, O', were used, the image would correspond to a projection of the object on to a plane perpendicular to the incident beam - as described in Sections IV.4 and VIII.4. Such a projection would superimpose information from all points above and below A, along the direction of the incident beam, throughout the thickness, t, of the specimen. But the diffraction pattern is not a plane section of Q-space and so the image does not correspond to a magnified projection of the object. Instead the microscope magnifies a selected section of the object eg. through A and B. Most of the material within the thickness t does not appear in the image - the microscope is said to possess a limited depth of field.

XII.3. Atomic resolution

Suppose that we wished to form an image with a resolution of about 0.1 nm, in order to determine the positions of atoms in an object. According to Section XII.2 we would need a source of waves whose wavelength, λ, was comparable with this required resolution. Three kinds of waves can have λ values of around 0.1 nm: electrons (Section II.6), thermal neutrons (Section II.6) and X-rays (Section I.2). Electron microscopy has three drawbacks. One: only very thin objects will not absorb the electrons - see Section VI.2. Macroscopic specimens have to be dispersed into tiny fragments or sectioned to produce objects which are about 100 nm thick. Both techniques, but particularly sectioning, involve extensive chemical and/or mechanical treatment which may produce an object which bears little resemblance to the original specimen. Two: elements of low atomic number scatter insufficient electrons to provide much contrast in the image. This problem can be

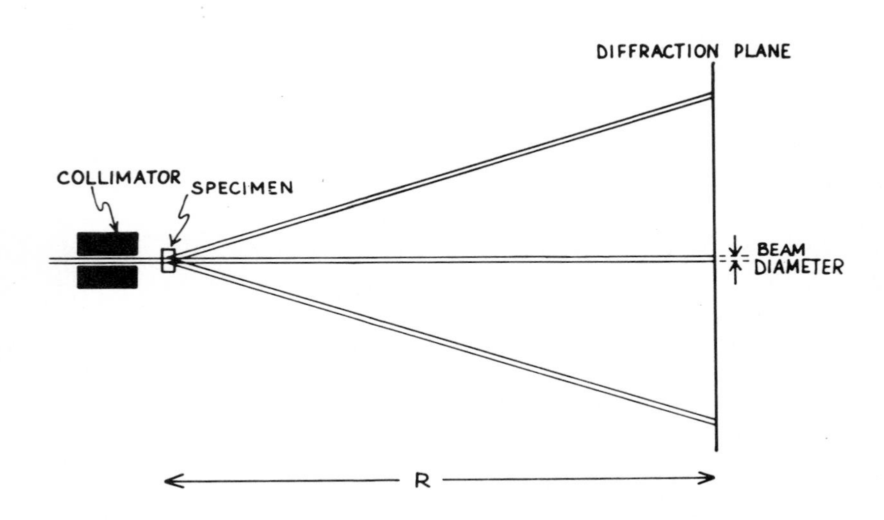

Fig. XII.2. Dispensing with an objective lens for recording a diffraction pattern when R is large.

overcome by attaching atoms which scatter strongly to the object - but this "staining" will change the original arrangement of atoms. Three: the object must be surrounded by a vacuum in the microscope, because air scatters appreciably (a very low pressure of around 10^{-4} N m^{-2} is required to produce a suitable beam). Thus objects which contain liquids cannot be examined; the liquid would evaporate, disrupting the structure of the object and ruining the vacuum. Neutron and X-ray microscopy are impossible because of the absence of lenses.

Some of the statements in the previous paragraph ought, strictly, to be qualified; many readers will not be unduly concerned with these qualifications and may wish to omit this paragraph. Firstly, some progress is being made towards examining "wet" specimens in the electron microscope. The object is enclosed in a cell so that liquid cannot evaporate into the high vacuum of the microscope. Very high voltages (around 1 MV) are used, in special microscopes, to accelerate the electrons so that they acquire sufficient energy to pass through "windows" in the cell. Secondly, lenses are available for those X-rays which have sufficiently high λ values - so-called "soft" X-rays. However, these wavelengths are far too high for an X-ray microscope to resolve atomic positions in an object. Even for low resolution structure determination, the X-ray microscope has serious disadvantages arising from soft X-rays being very strongly absorbed.

Can we dispense with any of the lenses in Fig. XII.1? Although the rest of this section is concerned explicitly with X-rays, most of the arguments apply equally well to neutrons - neutron diffraction is a less common technique because few reactors are available as sources. A parallel beam of X-rays can be produced by the methods described in Section I.2 - so there is no need for a condenser lens. Furthermore no lens is required to form the diffraction pattern because the cross-sectional diameter of the beam, which is typically around 0.1 mm, is small compared with a typical specimen-to-film distance of several centimetres. Figure XII.2 then shows that waves scattered with the same $\underline{Q}$ values will all arrive at effectively the same point on a plane even in the absence of a lens. Note that, in reality, the definition of the points is better than this figure might suggest because it shows an exaggerated diameter for the beam cross-section. But without a lens we cannot form an image.

In the absence of a suitable X-ray lens, we are reduced to analysing a diffraction pattern of the object. The obvious method of analysis is to compute an inverse Fourier transform to obtain an image. Remember that the phase information is lost when the diffraction pattern is recorded. But if this information can be deduced, by one of the methods given in Section III.4, an image can be computed. A structural model obtained by X-ray diffraction then has the same status as an image observed in a microscope; in Section X.4 it was shown that structural models of this kind are most readily obtained, at atomic resolution, when the object is a crystal.

Analysis of X-ray diffraction patterns is not always so straightforward. In Section XII.2 we saw that spurious details can appear when a limited range of Q-space is used to form an image. Limiting the range of Q-space to obtain low resolution structural information, which is equivalent to reducing the objective aperture of a microscope, is common in X-ray diffraction; sometimes this limitation is dictated by the nature of the specimen because there is no detectable scatter approaching Q_{max} of equation XII.1. Care must then be taken not to attribute spurious details, which arise from examining an insufficient extent of Q-space, to structural features in the object.

Another complication arises because the trial-and-error approach, of Section III.5, may be necessary to determine structure. In this approach a trial model is developed in the absence of phase information. Development of a model in this way amounts to making implicit assumptions about the phases of the scattered waves. Of course these assumptions may have to be modified during the course of the structure determination; nevertheless initial phases have implicitly to be assumed. Structural models derived by the trial-and-error approach do not then have quite the same status as the images formed by a microscope. Several alternative models may be proposed. The best is the simplest to provide a satisfactory explanation of experimental observations - these observations should not be restricted to those arising from X-ray diffraction experiments.

XII.4. Electron microscopy and diffraction

Electron lenses can be used to focus electrons to form an image in the electron microscope; unfortunately electron lenses have severe aberrations so that the theoretical resolution of around 4 pm is not achieved by a real microscope. Because electrons are charged, an electron beam can be deflected by a magnetic field. An electron lens is an electromagnet which focuses a parallel beam of electrons to a point on a focal plane - just like an optical lens focuses light. The electron beam is produced by accelerating the electrons through a potential difference which has a typical value of 100 kV. According to equation II.17 of Section II.6, their wavelength is then 4 pm. From Section XII.2 we would expect the resolution of the microscope to be of the same order. Unfortunately lenses cannot be manufactured with sufficient accuracy for such a high resolution to be achieved in practice. With a suitable specimen, such as a thin film of gold, the resolution of contemporary microscopes is no better than about 0.1 nm.

Of course, if we have some prior knowledge of eg. the arrangement of atoms in a film of gold, we may be able to obtain more detailed information than this resolution might suggest. In Section X.4 we saw that X-ray crystallography usually only yields a low-resolution picture of the electron density in a protein molecule.

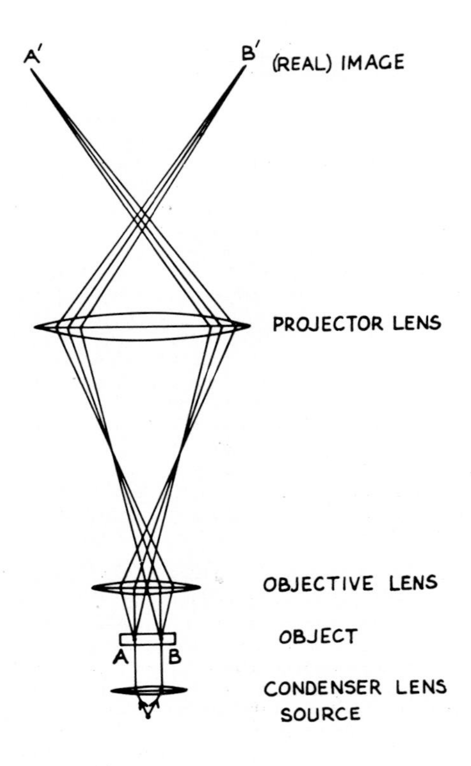

Fig. XII.3. Image formation in the projection microscope.

But, by combining this picture with stereochemical information, the positions of atoms in the molecule can be determined. Essentially the same approach can be used to obtain details of atomic positions in metals and minerals by electron microscopy.

Figure XII.3 shows how the image formed in the light microscope can be projected by replacing the eye-piece with a projector lens; a comparable optical arrangement has to be used in the electron microscope - the arrangement of Fig. XII.1 cannot be used because the eye is unable to detect electrons. The source, in an electron microscope, is a cathode. Electrons from the cathode are accelerated towards an anode by a high potential difference - many of them pass through a hole in the anode to produce a divergent electron beam. The beam can only be produced in a very high vacuum ie. a very low pressure - this vacuum is maintained throughout the instrument. A fluorescent screen allows the image to be visualised. When electrons strike a point on the screen, light is emitted; the intensity of this light depends on the number of electrons striking the point. Electron micrographs are recorded by replacing the fluorescent screen with a photographic film.

The image, produced by an electron microscope, corresponds to a magnified projection of the object ie. both the top and the bottom, as well as the inside, appear

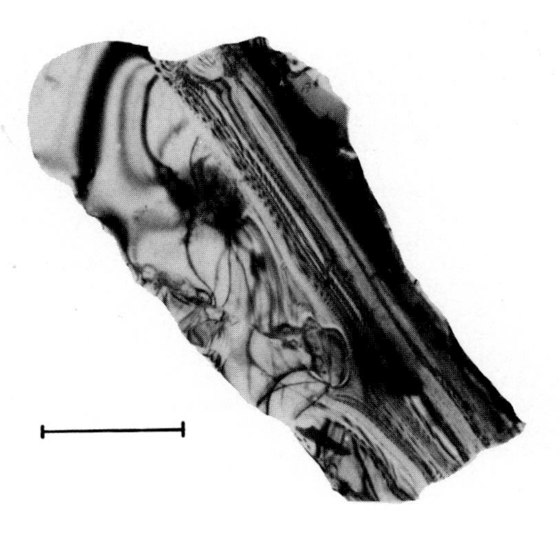

Fig. XII.4. Electron micrograph of a mica crystallite. The bar represents a distance of 10 μm (taken by S.M.W. Grundy).

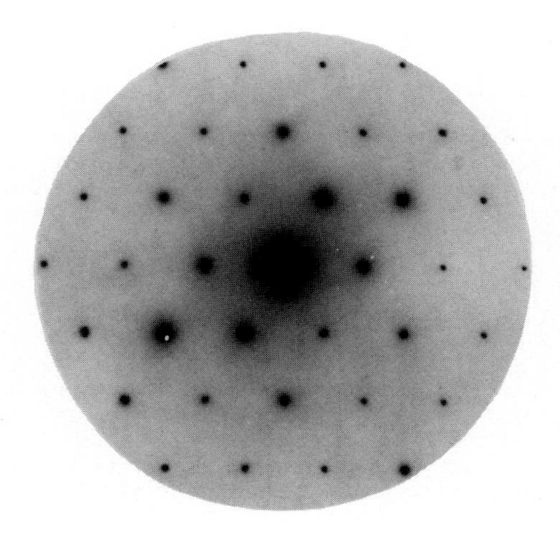

Fig. XII.5. Electron diffraction pattern of a part of the mica crystallite shown in Fig. XII.4 (taken by S.M.W. Grundy).

superimposed in the image. In other words the electron microscope has a considerable depth of field. This property arises because the microscope is always operated far from its theoretical limit of resolution. The Q values of the waves which form the diffraction pattern are then much less than Q_{max} of equation XII.1 ie. much less than $2\pi/\lambda$ - the radius of the Ewald sphere. At these low Q values the Ewald sphere is effectively a plane section of Q-space, passing through the origin O', which is perpendicular to the direction of the incident electron beam. According to Section XII.2, the image formed from this diffraction pattern is a magnified projection of the object on to a plane perpendicular to this beam direction.

Much smaller objects can be used in diffraction experiments with electrons than with X-rays. Figure XII.4 shows an electron micrograph of a single mica crystallite - in the original micrograph the magnification was about 2,500 times. Because an object of this size can be imaged in the electron microscope, we know when there is a single crystallite in the electron beam; this beam can have a diameter of only about 20 nm at the object. The object is placed on a supporting stage and can be moved into view by suitably small displacements because the stage is manipulated by the operator via a geared-down mechanical linkage. When the object is suitably positioned its diffraction pattern can be viewed instead of forming an image. How do electrons pass through the stage? They do not have to - the object is actually supported by a thin (20 nm) carbon film which spans a hole in a specimen grid which, in turn, spans an aperture in the stage.

Figure XII.5 is the electron diffraction pattern of a selected area of the object. This area could be selected by limiting the condenser aperture of the microscope. In Fig. XII.4 the crystallite is positioned so that one of its crystal axes points in the incident beam direction. Suppose we label this axis as the c-axis. Then the plane section of Q-space which was used to form Fig. XII.5 is the plane defined by the reciprocal lattice vectors $\underline{a}^*$ and $\underline{b}^*$ of Section X.2 - since $\underline{c}$ is perpendicular to both $\underline{a}^*$ and $\underline{b}^*$. In this example the electron diffraction pattern contains the same information as the precession method provides in X-ray crystallography - an undistorted section of the reciprocal lattice. But here an undistorted section of Q-space can be recorded from very much smaller specimens than are required for X-ray diffraction.

Actually the crystallite had to be carefully oriented to ensure that its c-axis pointed in the incident beam direction. This orientation was achieved by tilting the stage about two axes which were mutually perpendicular both to each other and to the direction of the incident beam. If the object is reoriented, by tilting the stage, a different projection appears in the image. By recording micrographs of objects at a variety of orientations it is possible to gain an idea of their three-dimensional structures.

The ability to form an image gives electron diffraction an advantage over X-ray diffraction. It is often simpler to appreciate certain kinds of information about an object from its diffraction pattern than from an image eg. information concerning periodicity - witness the relationship between the intensity distribution of the diffraction pattern and the autocorrelation function in Section III.3. The corresponding image, which can be formed when electrons are used, provides a valuable check on the interpretation of the diffraction pattern. It also provides a source of phase information. When the diffraction pattern is recorded all phase information is lost since the photographic film detects $I(\underline{Q})$ - as in Section III.2. The microscope can be considered to perform an inverse Fourier transform on $F(\underline{Q})$ to synthesise an image of the object. Reversing the process by Fourier transformation yields $F(\underline{Q})$ which conveys the phase information lost when the diffraction pattern was recorded. But note that the simple Born approximation of Section II.6 can break down for the conditions which occur in the electron microscope.

XII.5. Summary

Formation of a diffraction pattern is an intermediate stage in the formation of an image. In a microscope the diffraction pattern is formed in the back focal plane of the objective lens; the lens focuses all waves from a scattering object which have the same $\underline{Q}$ vectors at a point in this plane. However the diffraction pattern is not recorded so that the waves continue with their phase relationship undisturbed. Because the lens causes the paths of these waves to converge, an image is formed. In a microscope the eye-piece lens magnifies this intermediate image to form a greatly magnified final image.

No suitable lenses exist for X-rays - we can then record a diffraction pattern but not an image. The diffraction pattern can be recorded without a lens by making the distance from the scattering object to the detector sufficiently large. Since an image cannot be formed we have to record the diffraction pattern. The "phase problem" arises because the phase relationship between the scattered waves is then destroyed.

When deductive methods for determining structure are applied, the missing phase information can be measured and an image obtained by computing an inverse Fourier transform. Deductive methods can nearly always be applied to diffraction patterns from single crystals. Often, however, trial-and-error methods have to be used to determine structures.

Electromagnetic lenses are used to focus the electrons in an electron microscope. Either an electron diffraction pattern or an image can be recorded from an object. The image contains the phase information which is lost when the diffraction pattern is recorded. It also allows very small objects to be positioned in the electron beam so that diffraction experiments can be carried out with much smaller specimens than is possible with X-rays.

Bibliography

1. Introduction

This bibliography is intended to indicate a few sources of further information - especially those which I have found particularly useful myself. It is not intended to be comprehensive; in some sections I have been very selective because so many books are available. As far as possible I have recommended books which give a broad coverage of each topic - these books usually give references to the original literature. Review articles and research papers are included only where I consider that they are especially useful.

2. General

R.W. James, "The Optical Principles of the Diffraction of X-Rays", Bell, London (1954)

The best single source of information on the underlying physics of X-ray diffraction it is hardly concerned at all with applications. An excellent reference book since the material is covered in considerable depth.

A. Guinier, "X-Ray Diffraction in Crystals, Imperfect Crystals and Amorphous Bodies", Freeman, London (1966)

A useful source of information on X-ray diffraction theory; its approach is more rigorous than that adopted here and more topics are covered.

B.D. Cullity, "Elements of X-Ray Diffraction", 2nd ed., Addison-Wesley, London (1978)

Detailed accounts of many applications of X-ray diffraction - particularly the practical aspects. Bragg's law is the starting point from which the theory is derived; there is no mention of Fourier transforms.

"International Tables for X-Ray Crystallography", Kynoch Press, Birmingham (1969-74)

The four volumes of this book have a misleading title - only the first is exclusively concerned with diffraction by crystals. They are a mine of useful information on atomic scattering factors, scattering by dilute solutions of colloidal particles, diffraction geometry, Lorentz factors etc.

3. Chapter I

H. Semat, "Introduction to Atomic and Nuclear Physics", 4th ed., Chapman and Hall, London (1962)

Most books on atomic physics cover the material in Chapter I. Semat gives a particularly simple and readable account with references to sources of further information.

4. Chapter II

D.C. Champeney, "Fourier Transforms and their Physical Applications", Academic Press, London (1973)

The most useful account of Fourier transforms for supplementing this book; the appendices are a useful source of proofs.

R. Bracewell, "The Fourier Transform and its Applications", McGraw-Hill, London (1965)

Similar to the book by Champeney but with emphasis on applications in electrical engineering.

G.N. Watson, "A Treatise on the Theory of Bessel Functions", 2nd ed., Cambridge University Press, Cambridge (1966)

A detailed account of Bessel functions - useful as a reference book.

G. Harburn, C.A. Taylor and T.R. Welberry, "Atlas of Optical Transforms", Bell, London (1975)

Uses optical diffraction to illustrate many of the properties of Fourier transforms which are either stated or derived analytically in this book.

M. Kerker, "The Scattering of Light and other Electromagnetic Radiation", Academic Press, London (1969)

A very detailed account of exact and R.D.G. theories of light scattering. The formulae usually apply to unrationalised units; but sometimes formulae appropriate to SI units creep in without comment.

H.C. van de Hulst, "Light Scattering by Small Particles", Chapman and Hall, London (1957)

Covers rather less material than Kerker but provides more explanation. Exact theories are included - the "small" in the title only implies that the geometric optics approach is excluded.

G.E. Bacon, "Neutron Diffraction", 2nd ed., Clarendon Press, Oxford (1962)

A reasonably detailed, yet readable, account of neutron diffraction.

Electron diffraction and microscopy are covered in Section 14.

5. Chapter III

The material in this chapter is covered in most books on X-ray crystallography - see Section 12.

6. Chapter IV

C.A. Taylor and H. Lipson, "Optical Transforms : Their Preparation and Application to X-Ray Diffraction Problems", Bell, London (1964)

Gives an account of the methods used to simulate X-ray diffraction with light. More results from this technique are given in the book by Harburn et al. (Section 4).

Further information on diffraction geometry appears in Vol.II of the "International Tables for X-Ray Crystallography" (Section 2).

7. Chapter V

B.I. Bleaney and B. Bleaney, "Electricity and Magnetism", 2nd ed., Clarendon Press, Oxford (1965)

Gives a useful account of the intensity of electromagnetic waves; useful for making the simple approach adopted in this book more rigorous.

The material in this chapter is covered in most books on X-ray crystallography (Section 12) and in the book by James (Section 2).

8. Chapter VI

J.T. Randall, "The Diffraction of X-Rays and Electrons by Amorphous Solids, Liquids and Gases", Chapman and Hall, London (1934)

A clear account of the methods used to obtain diffraction patterns from gases and the formulae used to interpret them.

O. Kratky and I. Pilz, "Recent Advances and Applications of Diffuse X-Ray Small Angle Scattering on Biopolymers in Dilute Solutions", Quarterly Reviews of Biophysics, 5, 481-537 (1972)

Gives a simple account of the experimental methods used to measure the intensity of X-rays scattered by solutions of macromolecules and colloidal particles, as well as the theory used to interpret the results. A good source of references to the original literature.

The book by James (Section 2) is an especially useful source of further information.

9. Chapter VII

R. Hosemann, "Paracrystals in Biopolymers and Synthetic Polymers", Endeavour, 32, 99-105 (1973)

A simple account of the paracrystal theory of X-ray diffraction by non-crystalline systems.

C.A. Croxton, "Introduction to Liquid State Physics", Wiley, London (1975)

A comprehensible account of recent theories of the liquid state - including theories of the interference function.

The books by James (Section 2) and Randall (Section 8) are useful sources of further information on diffraction by liquids and amorphous solids.

10. Chapter VIII

H. Lipson and C.A. Taylor, "Fourier Transforms and X-Ray Diffraction", Bell, London (1958)

The results of Sections VIII.2 and VIII.3 are derived for the case of a perfect three-dimensional crystal.

Other topics are covered in the books by James and Guinier - see Section 2.

11. Chapter IX

A. Klug, F.H.C. Crick and H.W. Wyckoff, "Diffraction by Helical Structures", Acta Crystallographica, 11, 199-213 (1958)

A brief but complete account of the theory of diffraction by helices.

P-G. de Gennes, "The Physics of Liquid Crystals", Clarendon Press, Oxford (1974)

Gives a clear description of the various kinds of liquid crystals and a detailed account of their properties.

12. Chapter X

J. Pickworth Glusker and K.N. Trueblood, "Crystal Structure Analysis : A Primer", Oxford University Press, London (1972)

A simple account of the principles and methods involved in determining the structures of molecules in single crystals. Includes an extensive bibliography which should be consulted for details of other books on X-ray crystallography - the bibliography is not so reliable for diffraction by less ordered systems.

M.M. Woolfson, "An Introduction to X-Ray Crystallography", Cambridge University Press, Cambridge (1970)

Similar in scope to the book by Pickworth Glusker and Trueblood but less elementary. Gives excellent accounts of many aspects of the principles of X-ray crystallography - experimental methods are also included.

T.L. Blundell and L.N. Johnson, "Protein Crystallography", Academic Press, London (1976)

An account of the specialised techniques used to determine the structures of globular protein molecules in single crystals.

W.L. Bragg, "The Crystalline State", Bell, London (1966)

A survey of crystallography, first published in 1933, which has subsequently been reprinted.

13. Chapter XI

H. Lipson and H. Steeple, "Interpretation of X-Ray Powder Diffraction Patterns", Macmillan, London (1970)

A very complete and readable account of the applications of X-ray diffraction by crystalline powders.

S. Arnott, "Fiber Diffraction Analysis of Biopolymer Molecules", Transactions of the American Crystallographic Association, 9, 31-56 (1973)

Gives a brief account of diffraction by crystalline fibres and an up-to-date account of methods used to refine structural models. Includes details of diffraction by the different kinds of "statistical" crystal structures that can arise. Unfortunately it is published in a rather inaccessible journal.

14. Chapter XII

S.G. Lipson and H. Lipson, "Optical Physics", Cambridge University Press, Cambridge (1966)

The most useful optics text to supplement this book. Includes detailed accounts of Fraunhofer diffraction and its relationship to image formation.

C.J. Taylor and B.R. Pullan, "Computer and Optical Processing of Pictures", in "Non-Destructive Testing", R.S. Sharpe Ed., Vol.II, pp 65-87, Academic Press, London (1973)

Describes the formation of images and how the properties of the Fourier transform can be exploited to analyse them.

D.F. Parsons (Ed.), "Short Wavelength Microscopy", Annals of the New York Academy of Sciences, 306, 1-339 (1978)

A collection of papers describing some recent advances in electron and X-ray microscopy.

A.M. Glauert (Ed.), "Practical Methods in Electron Microscopy", North-Holland, Amsterdam (1972)

At present there are three volumes of this book - more are planned. So far they deal with the principles of electron microscopy, the operation of the microscope, specimen preparation, electron diffraction and some of the methods which can be used to analyse electron micrographs.

J.M. Cowley, "Diffraction Physics", North-Holland, Amsterdam (1975)

Detailed account of diffraction physics which is particularly strong on electron diffraction, including calculation when the simple Born approximation is inadequate.

Index